Charles H. Ralfe

Clinical Chemistry

an account of the analysis of blood, urine, morbid products, etc.

Charles H. Ralfe

Clinical Chemistry
an account of the analysis of blood, urine, morbid products, etc.

ISBN/EAN: 9783337390884

Printed in Europe, USA, Canada, Australia, Japan

Cover: Foto ©berggeist007 / pixelio.de

More available books at **www.hansebooks.com**

CLINICAL
CHEMISTRY.

AN ACCOUNT OF THE

ANALYSIS OF BLOOD, URINE, MORBID PRODUCTS, ETC.,
WITH AN EXPLANATION OF SOME OF THE
CHEMICAL CHANGES THAT OCCUR IN
THE BODY, IN DISEASE.

BY

CHARLES HENRY RALFE,

M.A., M.D., CANTAB.,

FELLOW OF THE ROYAL COLLEGE OF PHYSICIANS, LONDON; ASSISTANT
PHYSICIAN AT THE LONDON HOSPITAL; FORMERLY DEMONSTRATOR
OF PHYSIOLOGICAL CHEMISTRY IN THE MEDICAL SCHOOL OF
ST. GEORGE'S HOSPITAL.

ILLUSTRATED WITH 16 ENGRAVINGS.

HENRY C. LEA'S SON & CO,:
PHILADELPHIA, PA.

PREFACE.

THIS work, as its title implies, has been written for a practical purpose, viz., to furnish students and practitioners with a concise account of the best methods of examining chemically, abnormal blood, urine, morbid products, etc., at the bedside or in the hospital laboratory. It has been purposely made as simple as possible.

In spite of the disparagements of such eminent clinical teachers as Graves and Trousseau, chemistry has become more and more important to the physician as a means of elucidating many pathological conditions, or of determining the character of the morbid changes effected in tissues or secretions. Indeed, it is becoming more and more evident that we must eventually look to Chemistry for information with regard to the primary alterations that occur in fluids and tissues, and which are the first step in every disease. But even if original investigation in this direction is not engaged in, the student or practitioner will find that by making a chemical examination of the secretions, abnormal products, etc., whenever he has an opportunity, he gains such considerable insight into the nature of the morbid processes producing them, as enables him more effectually to comprehend their nature and modify their ill effects.

As few medical schools are now without physiological and chemical laboratories, in which students can obtain the necessary manipulative skill, I have not cumbered the text with minute directions with regard to apparatus, or with instructions for the conduct of such simple operations as weighing, evaporation, filtering, drying precipitates, etc., as the students to whom this work is addressed will have already gone through a course of practical training. Those, however, who have not done so will find the necessary instructions in my work, " Demonstrations in Physiological Chemistry," which I wrote for the purpose of instructing second-year students in the practical operations of the laboratory.

London, September, 1883.

CONTENTS.

—————

CLINICAL CHEMISTRY.

CHAPTER I.

SECTION A.—ORGANIC CONSTITUENTS OF THE ANIMAL
BODY.

1. Object of study.— It is proposed, in the present work, to confine our attention to those points of chemistry in so far as they relate to the study of the chemical phenomena concerned in effecting abnormal qualitative and quantitative changes in the constitution of the tissues and fluids of the animal body, and their practical bearing in relation to clinical medicine and pathology. Before, however, proceeding to the main object of study of this branch of science, it will be necessary to consider briefly the proximate and ultimate composition of the principles of the animal body, and the nature of the processes which produce the various decompositions and variations that are constantly occurring under normal conditions.

2. Chemical composition of the tissues and fluids.—If we submit a tissue or fluid to the action of heat, we find that with a moderate degree, 100° C., it loses weight and becomes dry; the loss of weight is due to removal of WATER; the dry residue represents the SOLID material present in the substance submitted to analysis. If the heat be now raised considerably, the dry residue chars or carbonises, showing the presence of ORGANIC matter ; if the heat be long continued this undergoes complete combustion, leaving an

B

ash which resists all further change on the application
of heat ; this ash, which consists of mineral salts, is
known as the INORGANIC residue. If, however, we
submit the tissue or fluid to a slightly more complex
analysis, we shall find that the organic and inorganic
residue consist of various substances, each of which
can be removed by appropriate means. For instance,
if, after having driven off the water, we treat the dry
residue with ether, we find that the etherial solution
will yield, on evaporation, a greasy residue, which
represents the FATS extracted from the tissue or fluid.
If, after the removal of the fatty matters, we treat
the original residue successively with boiling alcohol
and boiling water, we shall find, on evaporating the
alcoholic solution and the aqueous solution respectively, that each will give a residue containing substances soluble in these agents, and which have been
extracted from the original mass by their means.
These substances, which are various, are designated as
EXTRACTIVES, and consist chiefly of certain organic
substances such as urea, uric acid, kreatin, etc., and
the SOLUBLE salts of the inorganic constituents, whilst
to separate them from each other further means of
analysis have to be employed, and special tests applied to identify their characteristic reactions. After
the fatty matter and the alcoholic and aqueous extractives have been removed, there remains an elastic and
somewhat horny mass, which consists of PROTEID
material (albumin, fibrin, globulin, etc.), and which
chars on being burnt, leaving a residue of such salts of
the inorganic constituents which are INSOLUBLE in
boiling water, and some of the soluble salts not taken
up by extraction by water. By this elementary
analysis we have learnt, first, that the tissue or fluid
consists of water and solids; secondly, that the solids
consist of organic and inorganic substances, and that
these may be further divided : (1) The organic into

proteid substances, fatty matters, and extractives ; (2) the inorganic into soluble and insoluble saline constituents. The next step in the investigation is (1) to separate and distinguish the constituents present in each group from each other ; in the Proteids, the various albumins; in the Fats, the saponifiable fats, cholesterin, etc.; and in the Extractives, the urea, uric acid, etc., and the composition of the Soluble and Insoluble salts ; (2) to determine the nature, special characteristics, and ultimate chemical composition of each substance so separated.

3. **Composition and constitution of organic substances.**—There is no essential difference between organic and inorganic chemistry. Organic chemistry is simply the chemistry of carbon compounds, and accordingly we find that the organic principles we meet with in the animal body consist of carbon united in various proportions with hydrogen, oxygen, nitrogen, and some less abundant elements, such as sulphur, phosphorus, and iron.

4. **Nitrogenous and non-nitrogenous compounds.**—The carbon compounds, or organic principles, for purpose of convenience are divided into two distinct groups, viz. : (1) The non-nitrogenous, and (2) the nitrogenous ; those in which the element nitrogen is absent, and those in which it is present. Both these groups are represented by the principles which form the basis of the animal tissues and fluids, the first by the starchy, saccharine, and oleaginous principles, the second by the proteid or albuminous. These principles, which are usually distinguished by the term proximate, after fulfilling their purpose in the economy, undergo a series of changes (*metabolic*), are broken up and oxydised, and are finally reduced ; the former to carbonic acid and water, the latter to carbonic acid, water, and ammonium carbonate (urea). But before this final reduction is

reached, various intermediate products are produced ; thus, the oxydation of the non-nitrogenous principles yields lactic acid $C_3H_6O_3$, oxalic acid $C_2H_2O_4$, acetic acid $C_2H_4O_2$, formic acid CH_2O_2, and ultimately carbonic acid CH_2O_3. The nitrogenous group, in addition to the formation of products identical with the above, yield by oxydation a series of bodies, the lowest term of which is urea, the ammoniated form of carbonic acid ; thus, leucin $C_6H_{13}NO_2$, and kreatin $C_4H_9N_3O_2$, and perhaps uric acid $C_5H_4N_4O_3$, are recognised antecedents of urea CH_4N_2O.

5. **Classification of the compounds of carbon.**—These are (1) the compounds of carbon with hydrogen, or the *hydrocarbons ;* (2) the compounds of carbon with nitrogen.

(1) The **hydrocarbons.**—The classification of these bodies is based upon the atomicity of carbon, which, being a tetrad element, requires 4 atoms of hydrogen, or some other monad, for its full saturation, *i.e.,* to satisfy all its combining powers ; the fully satisfied hydrocarbon molecule will therefore be represented by the formula CH_4. Each additional atom of carbon requires, however, only two additional atoms of hydrogen to maintain the saturation, because a portion of the combining power of each carbon atom is employed in linking the carbon atoms together. The following diagram illustrates this important theory, which, it must be remembered, is applicable to all kinds of carbon compounds.

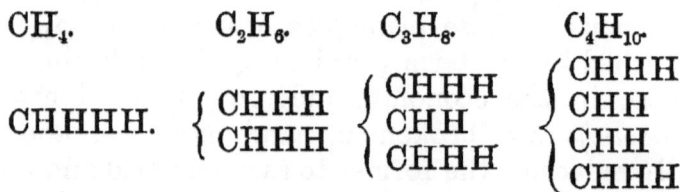

CH_4. C_2H_6. C_3H_8. C_4H_{10}.

CHHHH. $\left\{\begin{array}{l} \text{CHHH} \\ \text{CHHH} \end{array}\right.$ $\left\{\begin{array}{l} \text{CHHH} \\ \text{CHH} \\ \text{CHHH} \end{array}\right.$ $\left\{\begin{array}{l} \text{CHHH} \\ \text{CHH} \\ \text{CHH} \\ \text{CHHH} \end{array}\right.$

It follows from this that all carbon compounds

arrange themselves in series, the members of which differ from one another by CH_2 or by some multiple of CH_2. Thus we have formic acid CH_2O_2, acetic acid $C_2H_4O_2$, propionic acid $C_3H_6O_2$, and so on. Series of this kind are termed *homologous series*, and the members are said to be *homologues* of one another.

(A) *Hydrocarbon radicals.* From the above considerations, we deduce, as the general formula for a saturated hydrocarbon, the expression C_nH_{2n+2}, in which n may denote any number of atoms. If a hydrocarbon contains a less number of hydrogen atoms than the above formula requires, it is, or may be, a *radical;* that is, it may exist in compounds, and play therein the part of an elementary atom. Some hydrocarbons, however, which appear from their formulæ to be unsaturated are really saturated, or at any rate have an atomicity less than that indicated by the above theory. It is not necessary in this work to enter into the theoretical explanation of this apparent anomaly.

The *atomicity* of such a radical must obviously depend on the number of hydrogen, or other monad atoms required to complete it. Thus the radical CH_3 is a monad, CH_2 a diad, and CH a triad. Their function in compounds is well illustrated by their chlorides, which are strictly comparable to metallic chlorides.

Chloride of

sodium NaCl.	zinc $Zn''Cl_2$	bismuth $Bi'''Cl_3$
methyl CH_3Cl.	methylene CH_2Cl_2	formyl $CHCl_3$

Those of the hydrocarbon radicals which contain an *even* number of hydrogen atoms are capable of existing in the separate state, and most, but not all of them, have actually been prepared. Ethylene C_2H_4, and acetylene C_2H_2, are examples. Those, on the other hand, which contain an *uneven* number, such as methyl CH_3, ethyl C_2H_5, and glyceryl C_3H_5, cannot exist in the free state, but only in compounds.

The names and formulæ of a few of the more important hydrocarbon radicals are given in the following table.

TABLE OF PRINCIPAL HYDROCARBON RADICALS.

MONADS. Methyl series.	DIADS. Olefine series.	TRIADS. Glycerin series.
Methyl . . CH_3	$\overset{+}{}$	$\overset{+}{}$
Ethyl . . C_2H_5	Ethylene . C_2H_4	Ethine or Acety-lene . . . C_2H_2
Propyl . . C_3H_7	Propylene . C_3H_6	Propine or Ally-lene . . . C_3H_4
Butyl . . C_4H_9	Butylene . C_4H_8	Quartine or Cro-tonylene . . C_4H_6
Amyl . . C_5H_{11}	Amylene . C_5H_{10}	Quintine or Va-lerylene . . C_5H_8
Hexyl . . C_6H_{13}	Hexylene . C_6H_{12}	Sextine or Dial-lyl . . . . C_6H_{10}

(B) *Hydrocarbons of the aromatic series.*—*Benzene* C_6H_6, and *toluene* C_7H_8, are the most important members of this series. By the rule before given, they should be octads, but they possess the properties of saturated hydrocarbons, and are not therefore to be reckoned among the radicals. From them are derived the important monad radicals, *phenyl* C_6H_5, and *toluyl* C_7H_7.

The hydrocarbons, under various conditions of oxygenation and dehydration, form various well-known bodies, as alcohols, aldehydes, and organic acids.

I. *Alcohol.*—In an alcohol, a monad, diad, triad, etc., hydrocarbon radical replaces one or more atoms of H in one or more molecules of water. It is, in fact, a hydrate of a hydrocarbon radical, as the following show :

(a) Ordinary ethyl alcohol C_2H_6O is formed by the monatomic radical ethyl C_2H_5 replacing one atom of H in the single molecule of water thus, $\left.\begin{array}{c}C_2H_5\\ H\end{array}\right\}O.$

(b) Glycerin, or glyceryl alcohol $C_3H_8O_3$, is formed by the triatomic radical glyceryl C_3H_5''' replacing 3 atoms of H in the treble molecule of water $\left.\begin{array}{c}H_3\\ H_3\end{array}\right\}O_3$ thus, $\left.\begin{array}{c}C_3H_5'''\\ H_3\end{array}\right\}O_3.$

(c) Mannite $C_6H_{14}O_6$, a saturated hexatomic alcohol, is formed by the hexatomic radical C_6H_8 replacing 6 atoms of hydrogen in the molecule $\left.\begin{array}{c}H_6\\ H_6\end{array}\right\}O_6$ thus, $\left.\begin{array}{c}C_6H_8\\ H_6\end{array}\right\}O_6.$

II. *Aldehydes (alcohol de hydrogenatus)*.— If an alcohol be submitted to oxydation it loses two atoms of H, and is converted into a neutral body, called an *aldehyde*, which, having a great affinity for oxygen, rapidly absorbs it from the air, and is converted into an acid. Aldehydes are therefore compounds intermediate between the alcohols and the acids.

(a) Ethyl alcohol C_2H_6O deprived of 2 atoms of H forms *ethyl aldehyde* C_2H_4O, and ethyl aldehyde by oxydation yields *acetic acid* $C_2H_4O_2$.

(b) Mannite $C_6H_{14}O_6$ deprived of two atoms of H forms *mannitose* $C_6H_{12}O_6$, a sugar isomeric with glucose, dextrose, and other saccharine bodies, which are called *carbo-hydrates*. These, however, are not all aldehydes (some are ethers, others alcohols), but they are all derivatives of the hexatomic radicals. And mannitose by oxydation yields *mannitic* acid $C_6H_{12}O_7$. *Ketones* or *acetones* are bodies isomeric with the aldehydes, but are distinguished from them with regard to their behaviour with oxygen and hydrogen. With the former, an aldehyde unites directly to produce an acid ; with a ketone or acetone two acids are formed. With the latter, an aldehyde forms a primary alcohol, whilst with a ketone a secondary alcohol is produced.

III. *Organic acids.*— As stated above, the organic acids may be regarded as alcohols, in which a portion of the hydrogen of the radicals is replaced by oxygen. They are therefore formulated as derived from a single, double, or treble molecule of water by the replacement of H, H_2 or H_3 by a monad, diad, or triad oxygenated hydrocarbon radical.

Examples :

(*a*) Acetic acid $C_2H_4O_2$ is formed by the oxydised radical acetyl C_2H_3O, which has replaced one atom of H in water, thus $\left. \begin{array}{l} C_2H_3O \\ H \end{array} \right\}$ O.

(*b*) Glycollic acid $C_2H_4O_3$ is a double molecule of water in which half the hydrogen is replaced by *glycollyl* $\left. \begin{array}{l} C_2H_2O \\ H_2 \end{array} \right\}$ O_2. In fact, it is ethylene alcohol $\left. \begin{array}{l} C_2H_4 \\ H_2 \end{array} \right\}$ O_2 in which two atoms of hydrogen of ethylene are replaced by one atom of oxygen.

(*c*) Oxalic acid $C_2H_2O_4$ is a double molecule of water in which half the hydrogen is replaced by *oxalyl* $\left. \begin{array}{l} C_2O_2 \\ H_2 \end{array} \right\}$ O_2. Here the whole of the hydrogen is replaced by oxygen. It will be seen that both these acids are related to ethylene alcohol as acetic acid is to ethyl alcohol.

Amines.—When a hydrocarbon radical replaces the typical hydrogen of the molecule $\left. \begin{array}{l} H \\ H \\ H \end{array} \right\}$ N, the resulting compound is called a primary, secondary, or tertiary *amine*, according as one, two, or three atoms of hydrogen are replaced. Thus, the following amines are obtained by the substitution of the hydrogen of ammonia by methyl.

Ammonia.	Methylamine.	Dimethylamine.	Trimethylamine.
$\left.\begin{array}{l} H \\ H \\ H \end{array}\right\} N.$	$\left.\begin{array}{l} CH_3 \\ H \\ H \end{array}\right\} N.$	$\left.\begin{array}{l} CH_3 \\ CH_3 \\ H \end{array}\right\} N.$	$\left.\begin{array}{l} CH_3 \\ CH_3 \\ CH_3 \end{array}\right\} N.$
	Primary amine.	Secondary amine.	Tertiary amine.

Amides.—When an acid radical replaces any part of the typical hydrogen of ammonia, the resulting compound is called an *amide*. As some of the amides play a very important part in the animal economy, it is necessary to study their constitution a little more closely.

For this purpose it will be convenient to write the formulæ of a few important acids in a form which is a slight variation of that previously used.

$$C_2H_3O \quad HO \qquad \text{Acetic acid} \quad = \quad \left.\begin{array}{l} C_2H_3O \\ H \end{array}\right\} O$$

$$C_7HO_5 \quad HO \qquad \text{Benzoic ,,} \quad = \quad \left.\begin{array}{l} C_7H_5O \\ H \end{array}\right\} O$$

$$C_2H_2O \quad \overset{-}{H}O \quad \overset{+}{H}O \qquad \text{Glycollic ,.} \quad = \quad \left.\begin{array}{l} C_2H_2O \\ H_2 \end{array}\right\} O_2$$

$$C_2O_2 \quad \overset{+}{H}O \quad \overset{\perp}{H}O \qquad \text{Oxalic ,,} \quad = \quad \left.\begin{array}{l} C_2O_2 \\ H_2 \end{array}\right\} O_2$$

$$C_2H_4O \quad \overset{-}{H}O \quad \overset{+}{H}O \qquad \text{Lactic ,,} \quad = \quad \left.\begin{array}{l} C_3H_4O \\ H_2. \end{array}\right\} O_2$$

$$C_3O_3 \quad \overset{+}{H}O \quad \overset{+}{H}O \qquad \text{Mesoxalic ,,} \quad = \quad \left.\begin{array}{l} C_3O_3 \\ H_2 \end{array}\right\} O_2$$

In the above table we have acids of three different kinds, all capable of yielding amides.

(1) In acetic and benzoic acids we have examples of acids which are simply monobasic. The amides of these acids are called monamides, and are very simple :

Acetamide.

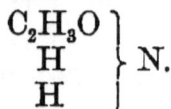

Benzamide.

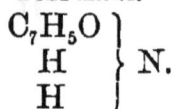

They differ from the corresponding amines, just as the acids do from the alcohols.

(2) Oxalic and mesoxalic acids are examples of dibasic acids. They are, in fact, dihydrates of the radicals C_2O_2 and C_3O_3. Now these radicals, being diads, are capable of replacing two atoms of hydrogen in the double molecule of ammonia. In this way neutral amides of the kind called *diamides* are formed.

Urea is by far the most important of the diamides.

Oxamide
(oxalyl diamide).

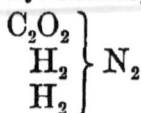

Urea
(carbonyl diamide).

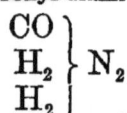

But from all dibasic acids a *monad* as well as a diad radical may be derived by merely deducting HO. Thus from sulphuric acid, SO_2 HO HO, we get not only the diad radical SO_2, but also the monad radical SO_2 HO. The following formulæ exhibit this :

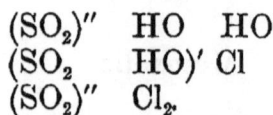

When one of these *monatomic* radicals of a *dibasic* acid replaces the hydrogen of ammonia, a monamide is formed. But, as there is still one atom of replaceable hydrogen attached to the radical, this atom can at any time be replaced by a metal or a hydrocarbon radical, and thus the acid character is not entirely lost. From a dibasic, the acid becomes, in fact, a monobasic one. Acids of this kind are called *amic acids*. Thus we have

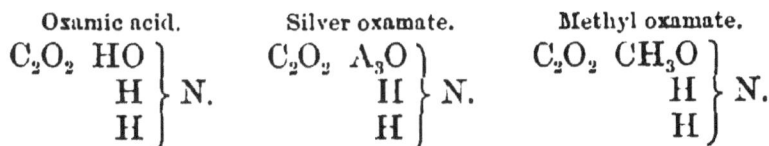

Oxamic acid.	Silver oxamate.	Methyl oxamate.
C_2O_2 HO $\left.\begin{array}{c} \\ \text{H} \\ \text{H} \end{array}\right\}$ N.	C_2O_2 A_3O $\left.\begin{array}{c} \\ \text{H} \\ \text{H} \end{array}\right\}$ N.	C_2O_2 CH_3O $\left.\begin{array}{c} \\ \text{H} \\ \text{H} \end{array}\right\}$ N.

(3) Glycollic and lactic acids are examples of acids which are *diatomic* as to their structure and *monobasic* as to their properties. Only one of the two typical hydrogen atoms that each contains can be replaced by metals. The difference has been indicated in the formulæ given above for the acids by marking the replaceable hydrogen by a *plus*, and the non-replaceable by a *minus*, sign. The effect of this peculiarity is that *two monad* radicals, one neutral and one acid, can be derived from each acid. These radicals replace one atom of hydrogen in the single molecule of ammonia, just as the monad radicals of dibasic acids do, and monamides are formed; but these monamides are amic acids if the radical so introduced contain the replaceable atom of hydrogen, or neutral amides if it contain only the non-replaceable atom. Thus from glycollic acid we have

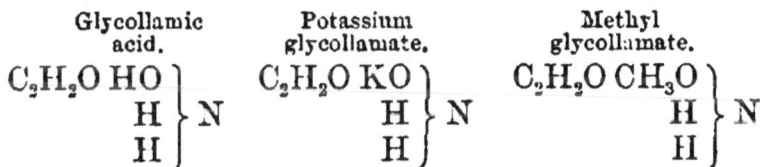

Glycollamic acid.	Potassium glycollamate.	Methyl glycollamate.
C_2H_2O HO $\left.\begin{array}{c} \\ \text{H} \\ \text{H} \end{array}\right\}$ N	C_2H_2O KO $\left.\begin{array}{c} \\ \text{H} \\ \text{H} \end{array}\right\}$ N	C_2H_2O CH_3O $\left.\begin{array}{c} \\ \text{H} \\ \text{H} \end{array}\right\}$ N

Glycollamide.

$$\text{and also } C_2H_2O \left.\begin{array}{l} HO \\ H \\ H \end{array}\right\} N.$$

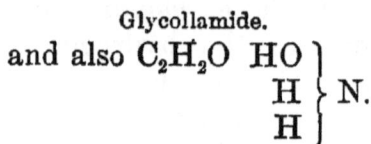

Analogous to this last we have *leucin*, the neutral amide of leucic acid, which is one of the homologues of glycollic acid.

Derivations of urea. — A somewhat numerous and complex class of bodies is known, the members of which contain radical, or *residues* of urea, together with radicals derived from various acids. Many of the derivatives of uric acid belong to this class, as also do the important compounds *kreatin* and *kreatinin.*

In many cases great difference of opinion exists as to the exact structure of these compounds, as they may be described by several formulæ. Kreatin is generally considered as containing residues of urea and sarcosin (methyl-glycocin). Its formula on the ammonia type may therefore be written as follows :

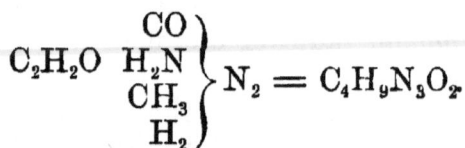

$$C_2H_2O \left.\begin{array}{l} CO \\ H_2N \\ CH_3 \\ H_2 \end{array}\right\} N_2 = C_4H_9N_3O_2.$$

It is capable of taking up the elements of water and splitting into urea and sarcosin. The following comparison of the formulæ of these two compounds will serve to illustrate this. The atoms which have to be removed to produce kreatin are printed in italics.

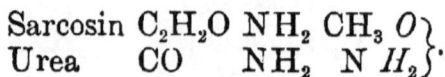

$$\left.\begin{array}{llll} \text{Sarcosin } C_2H_2O & NH_2 & CH_3 & O \\ \text{Urea} \quad\quad CO & NH_2 & N & H_2 \end{array}\right\}.$$

Compounds which contain one urea radical are

called *monureides.* Kreatin is a monureide, and so are *paraban* and *alloxan,* which are obtained by the oxydation of uric acid.

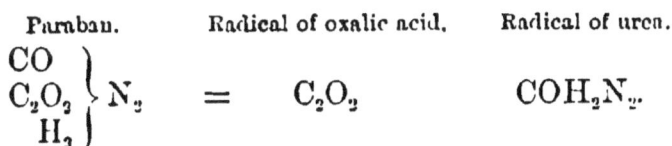

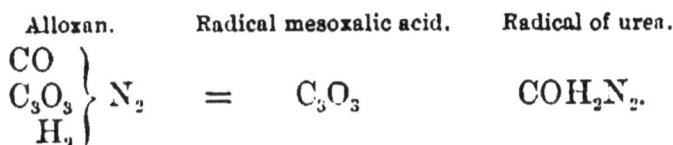

Paraban.   Radical of oxalic acid.   Radical of urea.

$$\left.\begin{array}{l} CO \\ C_2O_2 \\ H_2 \end{array}\right\} N_2 \quad = \quad C_2O_2 \qquad\qquad COH_2N_2.$$

Alloxan.   Radical mesoxalic acid.   Radical of urea.

$$\left.\begin{array}{l} CO \\ C_3O_3 \\ H_2 \end{array}\right\} N_2 \quad = \quad C_3O_3 \qquad\qquad COH_2N_2.$$

Compounds which contain two urea radicals are called diureides. *Allantoin, xanthin, hypoxanthin,* and *uric acid* belong to this class. There is still some doubt as to the exact constitution of uric acid. It is best represented by the hypothetical formula as consisting of one radical of tartronic acid and two of urea.

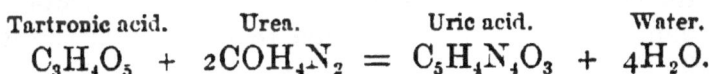

Tartronic acid.   Urea.   Uric acid.   Water.

$$C_3H_4O_5 \ + \ 2COH_4N_2 \ = \ C_5H_4N_4O_3 \ + \ 4H_2O.$$

(2) **Compounds of carbon with nitrogen.—** Carbon and nitrogen do not directly unite, but there exists a series of compounds containing the monatomic radical CN. called cyanogen. These cyanogen compounds may be regarded as derivations of ammonia, and are also connected with the compounds formed by oxalic acid with ammonia. Thus, we saw, when considering the *amides,* the diad radical of oxalic acid was capable of replacing two atoms of hydrogen in the double molecule of ammonia, forming a neutral amide of the kind called *diamide;* thus, urea also belongs to this group of *diamides,* since in it the diad

radical CO″ replaces two atoms of hydrogen in a double molecule of ammonia.

Oxamide (oxalyl diamide).

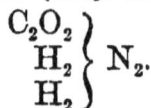

$$\left.\begin{array}{l} C_2O_2 \\ H_2 \\ H_2 \end{array}\right\} N_2.$$

Urea, however, can be formed directly by heating ammonium cyanate; thus,

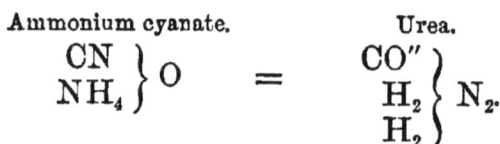

Ammonium cyanate.   Urea.

$$\left.\begin{array}{l} CN \\ NH_4 \end{array}\right\} O \quad = \quad \left.\begin{array}{l} CO'' \\ H_2 \\ H_2 \end{array}\right\} N_2.$$

Other compounds of carbon with nitrogen exist, whose exact constitution are not yet precisely determined. (a) *Alkaloids :* nitrogenous bases believed to belong to the compound ammonias; they combine with acids to form salts, and double crystallisable salts with platinic chloride. They are derived from the vegetable kingdom, therefore their presence in the animal tissues and fluids when found in them is only incidental. A class of bodies derived apparently from putrefactive changes of animal substances called *ptoamines* are closely related to these vegetable alkaloids, and may be mistaken for them, especially for strychnia. (β) Certain *Colouring* matters, of which the indigo group is the chief. Indol C_8H_7N, one of the products of pancreatic digestion, stands at the head of this series, of which uro-xanthin or indigogen, one of the urinary pigments, and *indican*, a glucoside of indigo, are members. (γ) *Albuminous* or *Proteid* substances, such as fibrin, casein, globulin, egg albumin, etc., which form the basis of the tissues and fluids of the body. Some chemists hold that the proteids are formed by the combination of an azotised principle, of a dibasic acid character, with different saline bases in varying proportions. Others, that these substances contain a radical called *protein*, combined with more or less hydrogen, sulphur, and phosphorus, according to the nature of the substance.

6. **Synthesis and analysis.**—For many years it was supposed that organic substances could be formed only by the agency of a living organism. In 1828, however, Wöhler obtained urea by evaporating ammonium cyanate, and since that time chemists have obtained by artificial means a large number of compounds formerly obtainable only from animal or vegetable organisms. These syntheses are effected either by bringing together molecules of simpler constitution to form a more complex body, as in the case of hippuric acid;

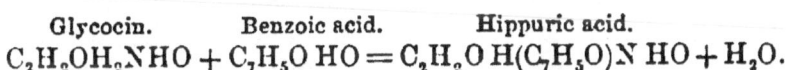

Glycocin.   Benzoic acid.   Hippuric acid.

$$C_2H_2OH_2NHO + C_7H_5O\ HO = C_2H_2O\ H(C_7H_5O)N\ HO + H_2O.$$

or by building up an organic compound from purely inorganic sources; as Berthelot obtained formic acid by heating carbon monoxide with potassium hydrate at 100° C.

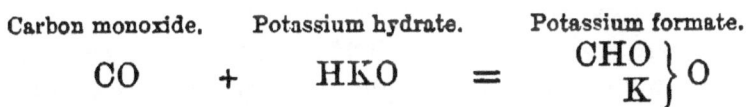

Carbon monoxide.   Potassium hydrate.   Potassium formate.

$$CO\quad +\quad HKO\quad =\quad \left.\begin{matrix} CHO \\ K \end{matrix}\right\}O$$

or, as Kolbe formed acetic acid from carbon disulphide.

In nature this formation of organic compounds from inorganic materials (*synthesis*) is effected by the agency of the vegetable kingdom. The plant under the influence of the rays of the sun liberates a quantity of oxygen from inorganic constituents such as carbonic acid, water, and ammonium carbonate,

which exist in the soil and air, converting them into
those saccharine, oleaginous, and albuminous principles which form its tissues and juices, and which
ultimately furnish the animal world with food. For
example, carbonic acid by deoxydation under certain
conditions may yield mannite ; thus

$$6CO_2 + 7H_2O - 13O = C_6H_{14}O_6.$$

That plants exert this deoxydising power is readily
shown by a very simple experiment. If a bunch
of fresh green leaves be plunged into a broad necked
bottle containing fresh spring-water, or water containing carbonic acid in solution, the bottle turned mouth
downwards into a basin of water so as to exclude the
air, and the whole placed in strong sunlight for an
hour or more, the leaves will become covered with
minute bubbles of oxygen, which is derived from the
decomposition of the carbonic acid of the water ;
the oxygen being set free, while the carbon is
absorbed by that plant to form its tissues. In this
process of deoxydation, however, a considerable
quantity of force derived from the sun's rays is
rendered latent, or, to speak more accurately, becomes potential, one portion being taken up by the
liberated oxygen, the other accumulated in the
tissues and juices of the plant; and this force will
remain latent till the oxygen and carbon are again
united.

The reunion of carbon with oxygen is effected,
either by the direct burning of carbon in oxygen,
as takes place when fuel is burnt in our grates or
stoves; or when the carbon elements of food and
tissues are submitted to the action of the respired
oxygen, and the potential energy assumes the active
or *kinetic* condition of heat and motion, whilst the
carbon compounds are reduced again to their original

inorganic state of carbonic acid, water, and ammonium carbonate. That animals exhale carbonic acid is demonstrated by the turbidity produced by passing a current of expired air through lime water, the lime being converted into calcium carbonate or chalk ; and we know that oxygen is absorbed, from the fact of its diminution in the atmosphere of close, crowded, and ill-ventilated apartments. Thus the vegetable organism is chiefly employed in building up *synthetically* inorganic into organic matter, whilst the animal *analytically* reduces organic compounds back again to their original inorganic constituents. These processes of deoxydation and oxydation do not at once raise or reduce the various substances to or from their elementary condition. On the contrary, the process is slow and gradual, several intermediate products being formed. De Luca, for example, has shown that in the ripening of the olive carbonic acid is first replaced by certain acids, as oxalic, tartaric, etc., and these by mannite ; which in its turn is deoxydised and converted into olein. On the other hand, in the decomposition of albumin a number of intermediary bases and fatty acids are formed, such as uric acid, xanthin, kreatin, lactic acid, and oxalic acid, before the final oxydation to urea and carbonic acid.

7. Oxydation and fermentation.—Although it may be broadly stated that the processes going on in the animal body are finally analytic, still certain synthetic processes do occur, as for instance the conversion of carbohydrates into fat, and the elevation of the proteids and peptones into tissues of a more complex form. These processes, however, have not as yet been sufficiently investigated. The downward transformation towards carbonic acid and urea, which non-nitrogenous and nitrogenous substances undergo, and to which the term *metabolism* has been applied,

c

have been the subject of much study and consideration, though our knowledge is still very imperfect in this direction. Formerly it was held that tissue changes depended on the amount of oxygen taken in by the lungs, so that on increased respiration a more intense combustion took place, and metabolism was increased with the production of more carbonic acid and urea, whilst, when respiration was impeded, oxydation was imperfectly performed, and, as a consequence, many of the intermediary products, as oxalic acid, uric acid, etc., were not burnt off, but were eliminated in an imperfectly oxydised condition. It is upon this view that most of the chemico-pathological speculations at present held are based. But the view is now gaining ground that the cells are to a certain extent independent of the amount of oxygen supplied to them by respiration; that is to say, though they originally obtain oxygen by the process of respiration, they are able, so to speak, to stow it away, and make use of it independently, under certain vital conditions which bring about intramolecular changes in their composition, so that reduction is a prior, or at least a simultaneous, process with oxydation. According to this view, instead of increased metabolism being the result of increased oxydation, it is the increase of the intramolecular action in the cells themselves that occasions the demand for oxygen, and a more active condition of circulation and respiration. Accordingly, in fever, the earliest step is the increase of intramolecular changes in the cells themselves, under the stimulus probably of the zymotic poison; for when the stored-up oxygen is exhausted, then a demand for a fresh supply causes an increased frequency of pulse and respiration, which continues so long as the stimulus (zymotic) acts on the cells and maintains this abnormal intramolecular activity. The fact that an increase in the amount of urea excreted by the urine

often precedes the rise of temperature* gives support
to this view, as does the fact of the gradual but steady
increase of pulse, respiration, and temperature, during
the early stages of febrile action.   For the acceptance of
this view it is necessary to discard the idea that oxydation occurs in the blood itself, and to hold that though
the quantity of hæmoglobin in the blood is the measure
of the oxydising power within the body, it is the
tissues that determine the amount of oxydation.   It is
not the place to discuss the physiological reasons and
experiments on which the view that the oxygen of
arterial blood passes into the tissues is based, and for
which the reader is referred to Professor Michael
Foster's work on "Animal Physiology;" it will be
sufficient to state that it is founded on the fact that
oxygen in the arterial blood reaches the tissues in a
state of high tension, while the oxygen in the tissues
is in a state of low tension.   As a consequence oxy-
hæmoglobin becomes reduced and passes on as venous
blood.

Among the processes which induce transformations
in complex organic substances, *fermentation*, next to
oxydation, holds an important place.

There are two kinds of ferments :

(*a*) *The organised ferments*, such as the yeast
plant, with powers of growth and reproduction,
and whose ferment power cannot be separated
from the ferment organism by filtration or any
solvent ;

(*b*) *The unorganised or the soluble ferments*, which
are freely soluble in water, and are incapable of
growth and reproduction.   These are the salivary,
gastric, and pancreatic ferments.

In the first group the action of the ferment is
towards the conversion of the substance into carbonic

* Ringer ; Med.-Chir. Trans. 1859.   Ralfe ; *Med. Times and
Gazette*, Jan., 1876.

acid, certain intermediate products being formed: thus, with

(1) *Yeast (Mycoderma cerevisiæ)*:

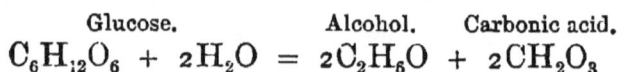

Glucose.   Alcohol.   Carbonic acid.

$$C_6H_{12}O_6 + 2H_2O = 2C_2H_6O + 2CH_2O_3$$

(2) *Lactic acid ferment (Bacterium lacticum)*:

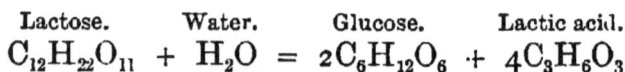

Lactose.   Water.   Glucose.   Lactic acid.

$$C_{12}H_{22}O_{11} + H_2O = 2C_6H_{12}O_6 + 4C_3H_6O_3$$

(3) *Butyric acid ferment (Baccillus subtilis)*:

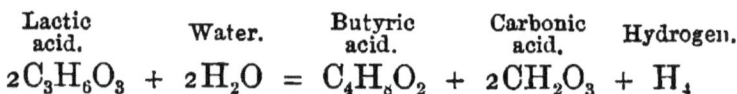

Lactic acid.   Water.   Butyric acid.   Carbonic acid.   Hydrogen.

$$2C_3H_6O_3 + 2H_2O = C_4H_8O_2 + 2CH_2O_3 + H_4$$

(4) *Urea fermentation (Micrococcus ureæ)*:

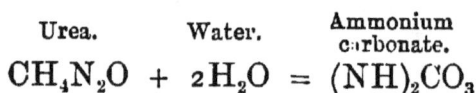

Urea.   Water.   Ammonium carbonate.

$$CH_4N_2O + 2H_2O = (NH)_2CO_3$$

In the second group, the process of fermentation is from anhydrides into hydrates. These ferments formed in the animal body have been called *enzymes*, and their action designated as *enzymosis*, and their nature as *enzymic.* They may be thus enumerated :

(1) *Ptyalin :*

Starch.   Water.   Dextrin.   Glucose.

$$2C_6H_{10}O_5 + H_2O = C_6H_{10}O_5 + C_6H_{12}O_6$$

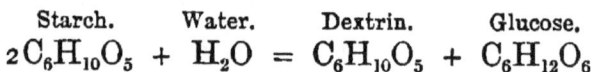

(2) *Pepsin pancreatin :*

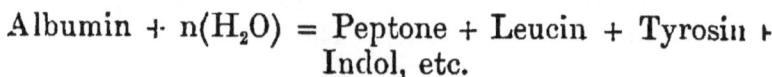

Albumin + n(H_2O) = Peptone + Leucin + Tyrosin + Indol, etc.

(3) *Pancreatic fat ferment :*

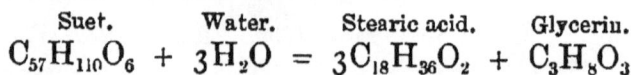

Suet.   Water.   Stearic acid.   Glycerin.

$$C_{57}H_{110}O_6 + 3H_2O = 3C_{18}H_{36}O_2 + C_3H_8O_3$$

As Ewald has aptly observed, and as we shall see when we come to the consideration of the changes taking place in the animal fluids, all *purely physiological* fermentations in the animal body correspond to the unorganised, all *pathological* to the organised ferments. As, for instance, the undue formation of acetic and lactic acids in the stomach and intestinal canal, and the decomposition of urea in the bladder into ammonium carbonate under the influence of the micrococcus ureæ. All ferments decompose peroxide of hydrogen and act best at temperatures between 20°— 70" C. At much higher, or at much lower, temperatures, the action is destroyed. For the growth of organised ferments it is necessary that they should be supplied with sufficient food, of which ammoniacal salts and alkaline phosphates are the chief.

SECTION B.—INORGANIC CONSTITUENTS OF THE ANIMAL BODY.

8. **Purposes of the inorganic constituents.** —The inorganic constituents subserve two important offices in the economy, viz. :—

(1) *Chemical.* — In effecting certain metamorphoses in the tissues and fluids, and keeping in solution many of the otherwise insoluble organic principles.

(2) *Mechanical.*—In giving strength and firmness to those textures, which, like bone, cartilage, and muscle, form the solid portion of the organism.

It must not be overlooked, however, that many inorganic substances are incidentally introduced into the body, and are eliminated without subserving apparently any purpose.

9. The **chemical rôle of the inorganic substances in the organism.**—The fact that, under normal conditions, the same weight of saline constituents are recoverable from the urine and fæces as are introduced during the same period with the food and drink, led physiologists and chemists for a long time to imagine that no change took place in their constitution during their passage through the body, and, consequently, to overlook the part played by the inorganic substances in histogenesis; or their influence in producing the daily and hourly variations which occur in the chemical composition of normal blood; or the action, physical as well as chemical, the inorganic constituents of each tissue have on the albumin, fats, water, etc., that compose that tissue, and how far excess or diminution of these constituents influences oxydation and nutrition going on in textures. Professor Parkes* was among the first to draw attention to the vastness and complicity of the chemical circulation taking place in the body, and the necessity for studying the chemical relations subsisting between the blood and the secretions. Attention once drawn to the subject, its importance was recognised and astonishment expressed that problems which so manifestly called for solution had been so long ignored. Although this branch of animal chemistry is the least developed, still our knowledge in respect to it is advancing. It has now been shown that whilst salts pass with immense and usually uniform rapidity into the circulation, and from thence to tissues, their discharge is by no means so regular, and they are detained for very unequal periods, which apparently depend on the need of the tissue to which they are supplied.

By feeding animals on food rich with acid salts,

* Gulstonian Lectures. *Med. Times and Gazette*, vol. i., p. 333, 1855.

Hoffmann[*] and Loscar[†] have shown, that however great the tendency of uric acid and of the acid salts of phosphoric acid is to combine with bases, yet these were not withdrawn from the alkaline blood, but were evidently withheld to maintain its alkalinity. These experimental facts are borne out by what seems to occur in scurvy. That disease is brought about by the prolonged withdrawal of the organic acids of vegetables and of recently killed meat, and their salts, from the dietary of those affected. These organic salts by oxydation yield alkaline carbonates. Now the alkaline carbonates are the salts chiefly concerned in maintaining the alkalescence of the blood, and it has been found that when these are cut off by the withdrawal of vegetable food, the alkaline phosphates are not excreted in the usual amount in the urine, but are apparently retained in the body to maintain the normal alkalescence of the blood, which has suffered by the withdrawal of the alkaline carbonates. Dr. Gaskell,[‡] too, has shown experimentally that a dilute alkaline solution acts upon the muscular tissue of the heart so as to produce a powerful contraction, whilst a dilute acid solution produces an opposite effect. Variations in the alkalinity of the blood, therefore, probably cause disturbances of the circulation and so effect a secondary chemical influence on nutrition as well as a direct one. Dr. Ringer,[§] from experiments on the action of the salts of potash soda and ammonia on the frog's heart, has shown that they have a very varying action, both as regards their power of influencing the frequency of contraction and the value of each beat.

[*] " Ueber der Uebergang von freien Säurer durch das alkalische Blut in den Harn ;" " Z. für Biologie," vii.
[†] " Zur alkalescence des Blutes ;" " Archiv für Physiologie," Pflüger. 1874.
[‡] " Journal of Physiology," vol. iii., No. 1. 1880.
[§] Med.-Chir. Trans., vol. lxv., p. 191. 1882.

Moreover, it has been shown that the various salts introduced into the body do not pass through unaltered ; probably all undergo some change. In the case of sodium chloride only four-fifths of that taken into the system passes out as such, the remaining fifth being decomposed into acid potassium phosphate. When chloride of calcium is taken by the mouth, nearly the whole of the lime is found in the fæces as a carbonate, whilst the whole of the chlorine is recoverable from the urine as chloride of potassium or sodium. Such decompositions help us to explain the seeming paradox that from the alkaline blood acid secretions are formed. Thus, in the case of the urine I have shown experimentally * that its acidity was the result of the decomposition between sodium or potassium bicarbonate and neutral sodium phosphate, two salts which exist in the blood ; thus,

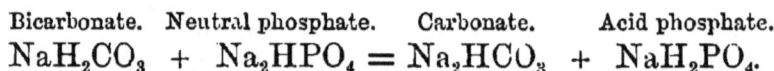

| Bicarbonate. | Neutral phosphate. | Carbonate. | Acid phosphate. |

$$NaH_2CO_3 + Na_2HPO_4 = Na_2HCO_3 + NaH_2PO_4.$$

And Maly,† who has investigated the subject of the acidity of the gastric juice with great care, has come to the conclusion that the hydrochloric acid is derived from the decomposition of neutral sodium phosphate with calcium chloride; thus,

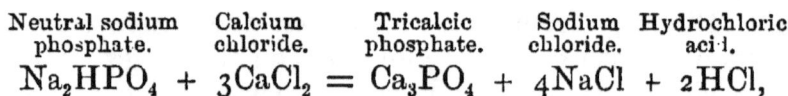

| Neutral sodium phosphate. | Calcium chloride. | Tricalcic phosphate. | Sodium chloride. | Hydrochloric aci·l. |

$$Na_2HPO_4 + 3CaCl_2 = Ca_3PO_4 + 4NaCl + 2HCl,$$

the acid in both instances diffusing out toward the free surface of the secreting membrane, the other salts remaining in the blood and returning to the circulation.

* *Lancet,* p. 29, July 4th, 1874.
† "Zeitschrift für Physiolog. Chimie," p. 174. 1877.

10. **Distribution of the inorganic constituents in the different tissues and fluids.—** The following table gives the percentage of inorganic residue in the principal tissues and fluids of the body. It must be remembered, however, with regard to the fluids, that the amounts are only approximate, since they vary greatly during the period of the twenty-four hours, under the different physiological conditions.

TABLE I.—PERCENTAGE OF INORGANIC RESIDUE IN

Enamel.	96·41	Pus.	0·84
Dentine.	71·30	Chyle	0 83
Bone.	66·70	Lymph	0·72
Muscle	1·82	Bile	0·78*
Nerve	1·74	Gastric juice.	0·24
Urine (24 hours)	1·32	Saliva	0·18
Blood plasma.	0·82	Pancreatic juice	—
Blood corpuscles.	0·72		

The different inorganic constituents are distributed in very varying proportions among the tissues and fluids ; thus, in muscle, in 100 parts of the ash the potassium salts are to the sodium salts as 58 to 23, whilst in blood they are as 6 to 79. Again, the distribution of the inorganic salts in blood is found to differ in the ash of the plasma and corpuscles relatively ; thus, in the ash of 1000 parts of corpuscles there is 3·679 parts of potassium chloride and 2·343 of potassium phosphate, whilst in the serum the potassium chloride only amounts to ·408 and a mere trace of potassium phosphate. On the other hand, the serum is particularly rich in sodium chloride whilst the corpuscles yield but little. Dr. Ringer's experiments, already alluded to, seem to throw some light on the reason for this marked divergence. The effect of the potassium salts on the action of the heart, says

---

* The bases combined with glycocholic and taurocholic acids not reckoned.

Dr. Ringer, was to increase contractility and excitability to a far greater extent than is the case with the
sodium salts, and therefore the former are to be
regarded as the most " poisonous " in their action, and
Dr. Ringer infers that "the action in one tissue being
selected and all other conditions being kept as far as
possible identical, if one salt prove itself more active
than another, it is at least not improbable that this
same salt will also prove itself more active under the
more complex conditions presented by the organism as
a whole." It can therefore be conceived that it is
advantageous that the sodium salts which are the least
active, as far as conditions of contractility and excitability are concerned, are those which remain free in
the serum, whilst the most active, the potassium
salts, are fixed, so to speak, in the corpuscles till
required to play their part in the nutrition of the
tissues.

11. The **conditions in which the inorganic constituents exist in the tissues and
fluids.**—The inorganic constituents enter and pass
out of the system as crystalloids, but whilst in contact
with organic matter they seem to lose their crystalline
form and become colloidal, and thus give the tissues
and fluids their homogeneous appearance. The influence of colloid media upon crystalline form has
recently received considerable attention, and a further
study will, no doubt, throw considerable light on the
pathological changes occurring in the solid tissues,
owing to morbid conditions of the colloidal medium
itself, or to an inadequate supply of the inorganic
constituents themselves, or to irregular distribution
(excess or deficiency) of the acids and bases. Attention was first drawn to the subject of molecular
coalescence by Professor Rainey,* and his work has

* " On the Mode of Formation of Shells, of Bone, and several
other Structures, by a Process of Molecular Coalescence." 1858.

been ably followed up by Dr. W. M. Ord,* Dr.
Vandyke Carter,† and Professor Harting.‡ The
following are some of the most important conclusions
arrived at. When two saline solutions, as, for instance, sodium carbonate and calcium chloride, which,
by double decomposition, are calculated to produce an
insoluble carbonate of lime, are allowed gradually and
slowly to intermix, through the intervention of a
viscous medium (such as dissolved gum or albumin),
there are formed by the union of nascent salt with
colloid, not crystals of the carbonate, but small, firm,
rounded bodies, which are possessed of a concentric
and radiate structure, and to which the term *sub-
morphous* is applied. These bodies, though disposed
to adhere to any surface, commonly remain free, but
also exhibit a tendency to meet and blend together, so
as to lead to the construction of a laminar series. The
chief conditions that influence the phenomena of
molecular coalescence may be thus enumerated : (1)
Nature of colloidal medium ; (2) nature of earthy
salts ; (3) temperature of the solution ; (4) density
of the solution ; (5) rapidity with which the saline
constituents intermingle. The converse of "molecular coalescence" is "molecular disintegration," which
Dr. Ord pronounces as the most strikingly original of
Professor Rainey's researches, and which promises to
bear much fruit in the elucidation of the nature of
certain bone diseases and the disintegration of renal
calculi. The inorganic substances occurring in the
animal tissues and fluids are not, for the most part,
united in a true chemical combination, but form, so
to speak, a loose physico-chemical combination, and
are, as it were, held in solution, from which

---

* "On the Influence of Colloids upon Crystalline Form and
Cohesion." 1879.

† "Mode of Formation of Urinary Calculi." 1873.

‡ "Artificial Production of some of the Principal Organic
Calcareous Formations." Utrecht. 1872.

they can be removed by mechanical means. Thus, for example, if some tissue, as muscle, be minced very fine and placed in a dialyser, and the dialyser floated in water, after the lapse of some hours the flesh will lose its consistence and become pulpy and jelly-like, at the same time the water will increase in density from the diffusion of the inorganic salts into it ; and these salts can be obtained for analysis by evaporating the diffusate. This process of diffusion is the best method for ascertaining the nature and chemical composition of salts as they exist in the tissues and fluids, since the result obtained by incineration is an artificial one, and does not represent the composition of the inorganic constituents as they exist in a natural condition in the tissues; since in burning off the organic matter the phosphorus and sulphur, which exist in proteid substances, become oxydised and form phosphates and sulphates. By applying the process of dialysis as a means of investigation to the exact constitution of inorganic constituents as they exist in the tissues, R. Maly * has arrived at very important results, one of the chief being the exact nature and composition of the saline constituents of the blood.

CHAPTER II.

Section A.—Enumeration of the Chief Organic Constituents of the Animal Body.

DIVISION I.—SACCHARINE AND STARCHY PRINCIPLES.

ALL contain six atoms (or multiples of six) of carbon, are generally known as carbohydrates. Berthôlet has shown that these substances are alcohols, or are related

* *Op. cit.*

to the alcohols, of the higher polyatomic radicals (§ 5). They closely resemble one another in their chemical characters and are isomeric, are neutral in their reaction, and have little disposition to enter into combination. They have all a strong action on polarised light. They are divisible into three groups.

Group I. Glucoses $n(C_6H_{12}O_6)$.

12. **Glucose** $C_6H_{12}O_6$ (syn. *dextrose, grape sugar*). —In pure state crystallises in rhombic tablets, but is usually met with in irregular warty masses. Soluble in own weight of cold water, the solution gives a dextro-rotatory power $+ 57.6°$. Undergoes *vinous* fermentation when yeast is added. Albuminous ferments induce *lactic* and subsequently butyric acid fermentation. Solutions of grape sugar become brown when boiled with liquor potassæ, and picric acid added to such solution gives a deep mahogany red. Alkaline solutions when heated reduce cupric salts, throwing down red precipitate of oxide of copper. An alkaline solution, heated with a few granules of bismuth, reduces the latter, and turns it black. A solution made faintly alkaline with sodic carbonate, and rendered blue by the addition of indigo, when heated to boiling, without agitation, becomes first violet, then yellow ; on agitation the blue colour is restored.

Lævulose $C_6H_{12}O_6$ (syn. *invert sugar*). — Incapable of crystallisation. Exists as a syrupy residue. Does not ferment so readily as grape sugar, but reduces copper from alkaline solutions readily. Differs from grape sugar in its left-handed polarisation, which diminishes as temperature rises ; being $- 106°$ at $15°$ C., $- 79.5°$ at $52°$ C., and $- 53°$ at $90°$ C.

Inosite $C_6H_{12}O_6 + 2H_2O$ (syn. *muscle sugar*).— Crystallises in two forms: (1) Large rhombic tables ; (2) small tufted groups of oblique prisms. Soluble in six parts of water at $20°$ C., insoluble in alcohol and ether, unfermentable with yeast, no action on polarised light; alkaline solutions do not reduce salts of copper, but give a greenish tint which clears up on standing, leaving original blue solution, but which again becomes green on heating. Heated to dryness on platinum foil, with a drop of nitric acid, the residue moistened with ammonia and calcium chloride yields a beautiful rose colour.

Group II. **Saccharoses** $n(C_{12}H_{22}O_{11})$.

13. **Saccharose** $C_{12}H_{22}O_{11}$ (syn. *cane sugar*).— Crystals, monoclinic prisms, very soluble in water, insoluble in absolute alcohol and water. Solutions have a dextro-rotatory power $+ 73·8°$. Boiled with water for some hours it is converted into a mixture of glucose and lævulose, or "*invert sugar.*" It ferments with yeast, but is transformed first into glucose. Does not at first reduce cupric salts from alkaline solutions, but does so after a while. In the intestines cane sugar is converted into "invert" sugar.

Lactose $C_{12}H_{22}O_{11} + H_2O$ (syn. *milk sugar*).— Crystals rhombic, soluble in six parts of cold water; dextro-rotatory power $= + 59·3°$. Does not readily undergo vinous fermentation; reduces copper in alkaline solutions. Boiled for some hours with dilute acids, forms *galactose*, this treated with nitric acid yields mucic acid.

Group III. **Amyloses** $n(C_6H_{10}O_5)$.

14. **Amylum** $C_6H_{10}O_5$ (syn. *starch*).—Granules, rounded, irregular form, marked with concentric laminæ, having a *hilum* or pore on surface. Insoluble in cold water, but when boiled swell up, burst, and form paste or mucilage. Solution is dextro-gyrous $+ 216°$. With iodine, starch solutions give a deep blue colour, which it loses when heated to 100° C., but regains it on cooling. Diastase, dilute sulphuric acid,

the salivary and pancreatic ferments, convert starch into glucose and dextrin. Tho further action of saliva is probably to convert the dextrin into glucose by the assumption of water, thus:

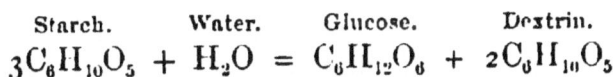

$$3C_6H_{10}O_5 + H_2O = C_6H_{12}O_6 + 2C_6H_{10}O_5$$

and

$$2C_6H_{10}O_5 + 2H_2O = 2C_6H_{12}O_6.$$

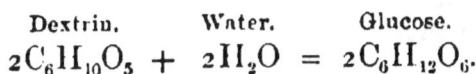

Dextrin $C_6H_{10}O_5$ (syn. *British gum*).—A yellowish powder, soluble in water, forming a viscous fluid. Solutions dextro-gyrous $= + 138\cdot8°$. With iodine gives reddish colour, which disappears on heating and does not reappear. Does not undergo vinous fermentation, or reduce copper salts till converted into glucose.

Glycogen $C_6H_{10}O_5$ (syn. *animal starch*).—Yellowish-white amorphous substance. Soluble in cold water, insoluble in alcohol. Readily converted into glucose. With iodine gives a similar coloration as dextrin, but is distinguished from it by the colour reappearing after it had been lost by heating.

DIVISION II.—THE FATTY PRINCIPLES.

The natural oils and fats existing in the body are all compounds of glycerin with fatty acids; the chief of which are mixtures of stearic, palmitic, and oleic acids. Thus, the tri-stearin of suet consists of three parts of the radical, stearyl $C_{18}H_{35}O$, which has replaced three atoms of typical hydrogen from glycerin C_3H_5 (OH)$_3$

thus, $\left.\begin{array}{c}C_3H_5 \\ 3(C_{18}H_{35}O)\end{array}\right\}O_3$. They are neutral bodies, of soft greasy consistence, highly inflammable. Insoluble in water. Soluble in ether, benzol, fluid oils, carbon

bisulphide, chloroform, and hot alcohol. Heated with
alkalies, they are *saponified;* that is, the fatty acid
unites with the alkali to form a soap, whilst the
glycerin is set free. Oils and melted fats, shaken up
with water containing albumin, bile, pancreatin, etc.,
become *emulsionised;* that is, the fatty matter is
broken up into small globules, which become more or
less permanently suspended in the aqueous solution.

15. **Stearin** $C_{57}H_{110}O_6$ (syn. *tri-stearin*).— The
chief constituent of solid fat. Occurs in white crystalline nodules. Melting point variable; 65° C. to
69·7° C. (Heintz). Soluble in boiling alcohol, from
which large square scales of stearin are deposited on
cooling.

16. **Palmitin** $C_{51}H_{98}O_6$ (syn. *tri-palmitin*).—
Fine needle-shaped crystals. Melting point variable;
mean 62·8° C. (Heintz). The substance known as
margarin consists of ten per cent. of stearin and
ninety per cent. of palmitin. *Margarin* is obtained
by heating fat in a water-bath, stirring with an equal
quantity of alcohol. The alcoholic solution on cooling
deposits needle-shaped crystals, arranged in whorled
groups or feathers. The melting-point of margarin is
lower than the melting-point of its two constituents,
being 47·8° C. (Heintz).

17. **Olein** $C_{57}H_{104}O_6$ (syn. *tri-olein*).— Colourless
oil, remaining liquid at 0° C.; exposed to air, it absorbs
oxygen, and becomes rancid. Heated to 280° C., it is
decomposed, and yields sebacic acid.

18. **Glycerin** $C_3H_5(OH)_3$.— Colourless syrupy
liquid. Soluble in water and in alcohol. Heated
with fatty acids, it combines with them, forming ethers
(glycerides) which constitute the neutral fats.

19. **Glycerin - phosphoric acid** $C_3H_9PO_6$.—
Syrupy liquid, both sour and sweet to taste. It is
dibasic, and its barium and calcium salts are soluble
in cold water, but not in alcohol. Its chief interest

is its association with a group of phosphorised bodies, of which lecithin is the chief.

DIVISION III.—PROTEID PRINCIPLES.

These constitute the basis of all the tissues of the body. They are amorphous, have low diffusive powers, turn the plane of polarisation to the left. Heated with caustic alkalies, they give off ammonia, volatile fatty acids as formic, acetic, etc., and yield leucin, tyrosin, and glycocin. Heated with strong nitric acid, they give a yellow colour, which turns orange on the addition of ammonia (*xantho-proteic reaction*). Boiled with mercuric nitrate solution, a red colour is developed (*Millon's reaction*). With acetic acid and ferrocyanide of potassium, proteids are precipitated from their solutions. Also by picric acid, tannic acid, or mercuric chloride. Acted upon by the gastric and pancreatic ferments they all become soluble, and acquire greater diffusive power.

Group I. **Native albumins.**—Soluble in pure water.

20. **Serum albumin.** — Viscid glairy fluid. Neutral in reaction, freely soluble in pure water. Its solutions have a specific rotatory power of $-56°$ C. When heated the solutions become opaque at $62·65°$ C., and coagulation occurs at $73°$ C. Strong mineral acids also produce coagulation, but they are not precipitated by sodium chloride, organic acids, nor dilute mineral acids.

Egg albumin differs from the above in that it is coagulated by ether, whilst serum albumin is not, and that the specific rotatory power for light is $-35·5°$ C., instead of $-56°$ C.

Group II. **Globulins.**—Not soluble in pure water, but in dilute neutral saline solutions.

21. **Globulin** (syn. *crystallin*).— As deposited in its coagulated form, by passing a current of carbonic

D

acid through its aqueous solution, globulin is insoluble
in pure water, but undergoes solution if the water is
saturated with oxygen. It is also soluble in dilute
neutral saline solutions. Solutions of globulin become
opalescent when heated to 73° C., and globulin is
deposited at 93° C. Globulin is precipitated from its
solutions by carbonic acid gas and by alcohol. (Constituent of aqueous and vitreous humours of eye.)

  Paraglobulin (syn. *serum globulin, fibrinoplastic*).
—As precipitated from blood serum by complete
saturation with magnesium sulphate, paraglobulin
is soluble in water saturated with oxygen, in dilute
neutral saline solutions (very dilute solution of
common salt, 0·5 per cent., precipitates paraglobulin
from solution; on addition of more salt the precipitate
redissolves, till about 20 per cent. is added, when
precipitate recurs), and in weak solutions of alkaline
carbonate, from which it is precipitated by alcohol.
Solutions coagulate at 75° C., but vary considerably,
according to amount of saline substance present in
solution. (Constituent of blood serum, and plasma,
colourless corpuscles of lymph and chyle.)

  Fibrinogen.—As obtained from blood by mixing
one-third its volume with saturated solution of magnesium sulphate, filtering and precipitating filtrate
with saturated solution of common salt; removing the
flaky precipitate, and frequently redissolving and re-
precipitating by alternately using solutions of common
salt of 7 per cent. and 20 per cent., to render it quite
free from paraglobulin. After the last precipitation,
dissolve in cold water, in which, owing to the salt
adhering to the precipitate, it is soluble. This solution coagulates at from 52° C. to 56° C. ; when serum
or a solution containing fibrin ferment is added, fibrin
is formed. (Constituents of blood serum, serous fluids,
and many pathological transudations.) The *fibrin
ferment* is prepared by adding to blood serum twenty

times its volume of alcohol, and allowing to digest
for a month or more. The insoluble matter is removed
and agitated with water and the mixture filtered.
The precipitate is then dried over sulphuric acid and
finely pulverised. The aqueous solution, added to a
solution containing fibrinoplastic and fibrinogen, causes
immediate coagulation.

Hæmoglobin.—Crystals of oxy-hæmoglobin are
rhombic plates or prisms with dihedral summits freely
soluble in water, insoluble in alcohol, ether, or chloroform. Solutions of oxy-hæmoglobin rendered sufficiently dilute, give two absorption bands in the spectrum in the yellow and beginning of green between D
and E, the band nearest D being the smaller and darker.
If to this solution we add ammoniacal stannous chloride, to which enough tartaric acid has been added to prevent precipitation, the two bands fade away, and there
appears a single broad band situated almost between
the two preceding ones. This is the band of reduced
hæmoglobin. Solutions of hæmoglobin readily decompose at temperatures above 0° C. ; on the addition of
acids and caustic alkalies, they break up into *hæmatin*
and globulin ; treated with glacial acetic acid and any
metallic chloride, it is decomposed into *hæmin*. Other
products of its decomposition are *hæmatoidin, hæmo-
chromogen, hæmatopopphyrin.*

22. *Myosin.*—As prepared by washing finely-
minced muscle with cold water till a precipitate is no
longer thrown down on the addition of mercuric chloride. The residue on filter is then treated with 10 per
cent. solution of sodium chloride, strained through
linen, the resulting liquid filtered and precipitated by
the addition of distilled water. In this state it is insoluble in pure water, but is so in very dilute saline
solutions (1 per cent.) ; from these it is precipitated
by the addition of common salt in bulk. Coagulates
at 70° C.

23. Group III. **Fibrin.**—Insoluble in water, and in dilute saline solutions. Does not dissolve in 1 per cent. solutions of hydrochloric acid, but swells up; pepsin added to this solution makes the fibrin soluble. Fibrin has the power of decomposing hydrogen peroxide, and giving a blue reaction with guiacum and etherial solution of hydrogen peroxide.

Group IV. **Modified albumins.**—Insoluble in water and dilute saline solutions, but soluble in dilute acids and alkali.

24. *Acid albumin (syntonin).*—Obtained by gradually heating a solution containing albumin with a dilute acid (1 per cent. solution of strong HlC), reprecipitated by neutralisation, but the precipitate is soluble in excess of the reagent. Its very dilute acid solution possesses a lævo-rotatory power of − 72°.

Alkali albumin (casein). — Obtained by heating solutions of albumin with dilute alkalies, reprecipitated on neutralisation, insoluble in excess, or in the presence of alkaline phosphates. The lævo-rotatory power of its alkaline solution prepared from sero-albumin is − 86°, from egg albumin − 47°.

25. Group V. **Peptones.**— Soluble in water, not coagulable by heat. Very diffusible, and pass easily through animal membranes. With an alkaline solution of cupric sulphate give a rosy red colour. Precipitated by picric acid, which is redissolved when warmed.

Group VI. **Albuminoids** or allied albumins, resembling in many points the proteids above described in chemical constitution, but exhibit in their characteristic reactions considerable differences among themselves. They occur in the epithelial and connective tissues.

26. **Mucin.**—Insoluble in cold water, but freely soluble in alkaline solutions, from which it is precipitated in strong masses by acetic acid, the precipitate not being dissolved by sodium sulphate; precipitated

by alcohol and alum, soluble in excess of the latter. Its solutions are not precipitated by heat or mercuric chloride, or potassium ferrocyanide and acetic acid.

27. **Gelatin.** — Insoluble in cold water, but freely soluble in hot, *gelatinising* when cold. It is not precipitated by acetic acid, but by mercuric chloride. Boiling with acids, or even prolonged boiling, prevents its warm solution *gelatinising* when cold.

28. **Chondrin.** — Soluble in hot water, gelatinising when cold, precipitated by acetic acid, but the precipitate is dissolved by sodium sulphate. Alum precipitates chondrin in excess.

29. **Elasticin.** — Highly insoluble, even at high temperature, its hot solution does not *gelatinise* on cooling, gives no precipitate with acetic acid.

30. **Lardacein.** — Insoluble in water and in dilute saline solutions. Not acted on by gastric juice. Gives a mahogany tinge when treated by iodine, and rosy red by methylanilin.

SECTION II.—PRODUCTS OF METABOLISM.

These are the products of oxydation of the non-nitrogenous and nitrogenous constituents, substances which enter into the composition of the body, and whose reactions have been discussed in the preceding sections of this chapter. The *non-nitrogenous* group consists chiefly of acids belonging to the fatty acid series, the aromatic groups and resinous acid, with some of their alcohols or aldehydes. The *nitrogenous* bases, or amides, are derived from the metabolism of the albuminous principles.

DIVISION I.—NON-NITROGENOUS.

Group I. **Fatty acids.**—Derived from the oxydation of the corresponding alcohols of the homologous series of hydrocarbon radicals, they are arranged in classes according as they are formed by monatomic or diatomic radicals.

Class I. **Monatomic fatty acids.**—The acids in this list are derived from the monatomic series of homologous hydrocarbons by the oxydation of the corresponding alcohols in which one atom of oxygen replaces two atoms of hydrogen; thus ethyl alcohol by oxydation loses two atoms of hydrogen, and is connected into aldehyde; and aldehyde by further oxydation becomes acetic acid ; thus,

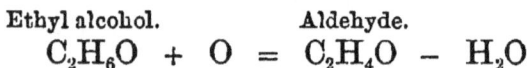

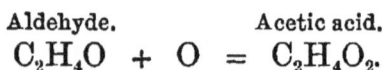

Ethyl alcohol.   Aldehyde.

$$C_2H_6O + O = C_2H_4O - H_2O$$

Aldehyde.   Acetic acid.

$$C_2H_4O + O = C_2H_4O_2.$$

31. **Formic acid** CH_2O_2.—Is a colourless corrosive liquid, boiling point 100° C., solid at 1° C.

32. **Acetic acid** $C_2H_4O_2$—Is a colourless sharp-smelling liquid ; boiling point 118° C., solid at 17° C.

Acetone.—Limpid, colourless liquid, sp. gr. 0·7921, with a peculiar etherial (decayed apple) odour. Solutions of acetone give violet-red, with ferric chloride ; with iodine and chlorine in the presence and alkalies it is converted into iodoform and chloroform.

Alcohol.—Heated with a few drops of sulphuric acid and solution of potassium dichromate, an emerald green colour is produced. Nitric acid added to alcohol, and this mixture warmed, gives off fumes of nitrous ether ; if to this a solution of mercurous nitrate be added and heat applied, a yellowish precipitate will be thrown down; this is mercuric fulminate. A solution of alcohol heated with potassium hydrate and iodine gives a yellow precipitate of iodoform.

33. **Propionic acid** $C_3H_6O_2$ — Is a colourless oily liquid; boiling point 140° C., solid at 20° C.

34. **Butyric acid** $C_4H_8O_2$.—Is a mobile colourless liquid; boiling point 162° C., solid at 20° C. ; odour of rancid butter; by fermentation lactic acid yields butyric acid, carbonic acid, and hydrogen.

35. **Valeric acid** $C_5H_{10}O_2$.—Is a limpid, colourless, oily liquid; boiling point 174° C., solid at 20° C.; odour of valerian.

36. **Caproic acid** $C_6H_{12}O_2$. — An oily liquid having the odour of acid sweat; boiling point 199° C., solid at 4° C.

37. **Capric acid** $C_8H_{16}O_2$.—Is a greasy oily liquid, which crystallises at 29° C. in colourless needles, which on heating evolve a goaty odour; boiling point 236° C.

38. **Palmitic acid** $C_{16}H_{32}O_2$— Is a tasteless, white, fatty substance; melting point 62° C., soluble in ether and alcohol, forming acid solutions which on concentration deposit white crystalline needles; insoluble in water. With glycerin it forms three bases (glycerides) : (1) mono-palmatin, (2) di-palmitin, and (3) tri-palmitin; the latter is a constituent of animal fat, which mixed with tri-stearin forms margarin.

39. **Stearic acid** $C_{18}H_{36}O_2$.— Is a white crystalline substance; melting point 69·2° C., soluble in ether and alcohol, insoluble in water. Like palmitic acid, it forms with glycerin three compounds, mono-stearin, di-stearin, and tri-stearin; the latter is a constituent of suet.

40. **Oleic acid** $C_{18}H_{34}O_2$.— Is solid at 4° C., liquid at 14° C. ; freely soluble in alcohol and ether, insoluble in water. By the action of nitrous acid it is converted into elaidic acid. By distillation it yields sebacic acid; this distinguishes it from other fatty acids. With glycerin it forms mon-olein, di-olein, and tri-olein; the latter forms the oily portion of the animal

fat.   Oleic acid is in reality derived from the glycerin series of homologous hydrocarbons, but for convenience is classified here.

Class II. **Diatomic fatty acids.**—These acids are derived from the "olefine series of homologous hydrocarbons" by the oxydation of the corresponding alcohols, and may be divided into two classes, viz., the *monobasic* acids which are formed by one atom of oxygen replacing two atoms of hydrogen in the corresponding alcohol; and the *dibasic* acids which are formed by the replacement of four atoms of hydrogen by two atoms of oxygen; thus ethylene alcohol by oxydation loses two atoms of hydrogen, and is converted into monobasic glycollic acid; and glycollic acid by further oxydation loses two atoms of hydrogen and becomes dibasic oxalic acid, thus :

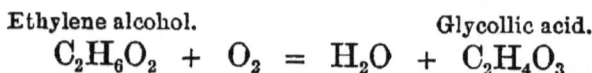

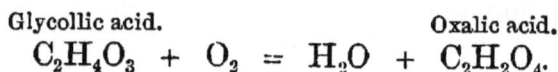

Ethylene alcohol.   Glycollic acid.
$$C_2H_6O_2 \ + \ O_2 \ = \ H_2O \ + \ C_2H_4O_3$$

Glycollic acid.   Oxalic acid.
$$C_2H_4O_3 \ + \ O_2 \ = \ H_2O \ + \ C_2H_2O_4.$$

SUB-CLASS (*a*)—MONOBASIC ACIDS.

41. **Carbonic acid** CH_2O_3. — Colourless inodorous gas.   Specific gravity 1·529 heavier than gas, soluble in an equal volume of water.   This solution reddens blue litmus paper, the red colour disappears on drying.   Carbonic acid gas passed through lime water produces a white precipitate of calcium carbonate.   It is best determined quantitatively by passing it through a solution of potassium hydrate and noting increase of weight in the solution, or by the process given for estimation of carbonates.

42. **Glycollic acid** $C_2H_4O_3$. — This substance does not exist in a free state in the organism.   Its ammoniated form, glycocin, conjugated with cholic

acid, forms the glycocholic acid of the bile; and with benzoic acid unites to form hippuric acid. Glycollic acid is a syrupy acid liquid, soluble in ether and alcohol, from concentrated solutions of which deliquescent crystals, which melt at 78° C., are deposited.

43. **Lactic acid** $C_3H_6O_3$.—Is a colourless syrupy fluid of sharp acid taste; specific gravity 1·21; soluble in water, alcohol, and ether. If distilled at temperatures above 160° C., it decomposes. Heated with sulphuric acid it evolves carbonic oxide. Heated with nitric acid it yields oxalic oxide. Its calcium and zinc salts· are characteristic. *Sarcolactic acid* (paralactic and ethylene lactic acids) closely resembles lactic acid and is isomeric with it; its salts, however, differ in crystallising with a smaller proportion of water, and in their crystalline form; it is obtained from muscular tissue.

46. **Leucic acid** $C_6H_{12}O_3$.—This acid only exists in the body in its ammoniated form, leucin, from which it can be obtained by heating with nitrous acid. (*See* Leucin.)

SUB-CLASS (b)—DIBASIC ACIDS.

47. **Oxalic acid** $C_2H_2O_4$. — Crystallises in prisms; its solutions have an intensely sour taste. Heated to 160° C., oxalic acid is partially decomposed into formic acid, carbonic acid, and water. With solutions of lime it forms the normal calcium oxalate, a highly insoluble salt, which is deposited from urine usually in the form of octohedral crystals of letter-envelope shape. Solutions of calcium oxalate are precipitated by the addition of alcohol. With silver nitrate they give a white precipitate, soluble in nitric acid and ammonia; if the precipitate be greatly heated on platinum foil it will decrepitate, leaving a residue of metallic silver.

48. Succinic acid $C_4H_6O_4$.—The crystals form large rhombic colourless tablets which fuse at 160° C., and are soluble in alcohol and cold water. These solutions give a brown precipitate with ferric chloride, and white with barium chloride. Found sometimes in hydrocele fluids and contents of ovarian cysts.

Group II.—Aromatic acid series.

49. Benzoic acid $C_7H_6O_2$.—Occurs in pearly white crystalline plates, which fuse at 121° C. Its solutions give reddish-brown precipitates with ferric chloride, and blue precipitates with cupric acetate. Found sometimes in stale human urine, is present in the fresh urine of herbivorous animals. Its chief interest is due to the presence of its radical in hippuric acid.

50. Carbolic acid C_6H_6O (syn. *phenol*).— Occurs as a white crystalline mass, melting at 42° C., and forming a heavy, oily, corrosive fluid with a pungent, smoky odour. It gives a violet colour to solutions of ferric chloride, which acquires a blue colour on exposure to the air; a chip of fir wood saturated with phenylic acid and dipped into dilute hydrochloric acid turns a deep blue colour. In carbolic acid poisoning the sulphates disappear from the urine, being converted into sulpho-carbolates. Allied to carbolic acid; *taurylic, damaluric,* and *damolic* acids are to be found in minute quantities in human urine.

Group III. **Resinous acids.**—The radical of these acids has not been isolated, but it is probable that it belongs to some of the higher aromatic hydrocarbons.

51. Cholesteric acid $C_8H_{10}O_5$.—Is formed whenever cholesterin is heated with nitric acid. It is a yellow, non-crystallisable substance, which rapidly absorbs moisture from the air; it is soluble in water, alcohol, and ether. *Cholesterin* $C_{26}H_{44}O$ may be regarded as an alcohol of this series, being homologous to cinnyl alcohol. As obtained from its hot alcoholic solutions it forms characteristic glistening rhombic plates with notched edges, which float on water; extremely soluble in ether. Heated with nitric acid it gives off yellow acid fumes of cholesteric acid; the residue touched with ammonia gives a red coloration. Cholesterin may be mistaken for leucin, but is distinguished from that body by its solubility in ether.

52. **Cholic acid** $C_{24}H_{40}O_5$. — Occurs in two forms, the amorphous and the crystalline. The former is resinous and viscous, very slightly soluble in water, but freely soluble in alcohol and caustic alkalies. In the latter the crystals are octohedral and tetrahedral; they are colourless, insoluble in water, very soluble in ether and alcohol; the octohedral variety contains 1 molecule, the tetrahedral $2\frac{1}{2}$ molecules of water; when heated they lose this water of crystallisation and become disintegrated. Cholic acid heated with acids at a temperature of 200° C. loses 2 atoms of water and is converted into *dyslysin* (so named from its insolubility in water), acids, alkalies, and alcohol. Cholic acid with a solution of sulphuric acid and sugar gives a deep purple coloration known as Pettenkoffer's test. Conjugated with glycocin and taurin it forms the bile acids *glycocholic* and *taurocholic* acids.

DIVISION II.—NITROGENOUS.

Group I. **Monamides.**

53. **Glycocin** $C_2H_5NO_2$, or amido-acetic acid. Crystals of glycocin are hard and granular, and have a sweet, mawkish taste; they are soluble in 400 parts of cold water, but quite insoluble in alcohol. A

transient fiery red colour is given when glycocin is
heated with a strong solution of caustic potash. A
stream of nitrous acid passed through an aqueous
solution of glycocin decomposes it, nitrogen is evolved,
and on agitating with ether and evaporating glycollic
acid is obtained. Glycocin is derived from the
decomposition of the gelatinous tissues; conjugated
with cholic acid it forms *glycocholic acid*, one of the
acids of the bile. This acid is deposited from its
alcoholic solution in long delicate colourless needles,
slightly soluble in cold water and ether, very soluble
in alcohol and boiling water. These solutions are
precipitated by *neutral* lead acetate; with strong
sulphuric acid and sugar they give a red coloration
(Pettenkoffer's test).

54. **Taurin** $C_2H_7NSO_3$, or ethyl amido-sul-
phuric acid, is prepared from bile. The crystals are
four-sided prisms with pyramidal extremities, insoluble
in ether and alcohol, soluble in 15 parts of cold water;
the aqueous solution has a neutral reaction. They are
dissolved by the mineral acids without change ; burned
in air they evolve sulphurous acid fumes; heated with
potash, ammonia is evolved, and potassium sulphate
formed. The aqueous solutions are not precipitated
by salts of mercury, copper, or silver. This substance
is found in the bile associated with cholic acid ; it
can be obtained from most glandular tissues, and from
the lung tissue and muscular fibre of the heart.
Taurocholic acid never occurs in a crystalline form,
but appears as an oily resinous fluid, of tawny colour,
very soluble in alcohol and ether, and has a strong
acid reaction; its aqueous solution, on heating, is
readily decomposed into taurin and cholic acid ; it
turns the plane of polarised light to the right, and
gives the purple reaction with sulphuric acid and cane
sugar. Its solutions are precipitated by *basic* lead
acetate.

55. Sarcosin $C_3H_7NO_2$, or methyl glycocin, contains one atom of glycollic acid radical $C_2H_3O_2$, and one atom of methyl CH_3 replacing two atoms of H. It is not met with in the animal body, its only interest being that it is a constituent of kreatin, that body yielding by decomposition both urea and sarcosin.

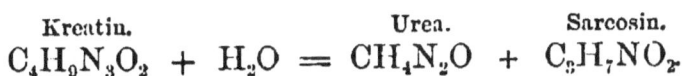

Kreatin.   Urea.   Sarcosin.

$$C_4H_9N_3O_2 + H_2O = CH_4N_2O + C_3H_7NO_2.$$

56. Cholin $C_5H_{15}NO_2$ (syn. *neurin*) may be regarded as ammonium hydrate in which 1 atom of H is replaced by 1 atom of oxy-ethyl C_2H_5O + 3 atoms of methyl $3CH_3$ making $\left. \begin{array}{c} C_2H_5O \\ (CH_3)_3 \end{array} \right\}$ NHO. Cholin is obtainable as a thick syrup, and has a powerful alkaline reaction; soluble in water and alcohol. Its solution prevents the coagulation of albumin. It does not exist free in the body, but is a constituent of lecithin, protagon, and cerebrin.

57. Cystin $C_8H_7NSO_2$, or amido-sulpho-pyruvic acid, an amide of the lactic acid series, pyruvic acid being lactic acid deprived of 2 atoms of hydrogen. Deposited from an ammoniacal solution, it forms hexagonal crystals, which have a tendency to overlap, and which acquire a yellowish-green tinge on exposure to air. They are soluble in alkalies and *strong* mineral acids. Boiled with caustic potash in the presence of lead acetate they yield a black precipitate of lead sulphide.

58. Leucin $C_6H_{13}NO_2$, or amido-caproic acid. —The leucic acid radical $C_6H_{11}O$ replacing one atom of H in ammonium hydrate; thus $\left. \begin{array}{c} C_6H_{11}O \\ NH_2 \end{array} \right\}$ O. Leucin is deposited from its hot alcoholic solution in white shining plates. Slightly soluble in cold alcohol; very insoluble in ether, which distinguishes it from cholesterin, which it closely resembles in appearance. Very soluble in boiling water. Leucin is interesting physiologically as being one of the antecedents of urea, and pathologically from its presence in the urine in certain diseases of the liver.

59. **Tyrosin** $C_9H_{11}NO_3$ is probably amido-propionic acid $C_3H_5(NH_2)O_2$ in which 1 atom of H is replaced by oxy-phenol C_6H_5O to make $\left. \begin{array}{l} C_3H_4 \\ C_6H_5O \\ NH_2 \end{array} \right\} O_2$.

The crystals are fine needles, which sometimes cluster to form stellate groups, or packs to form rounded balls. They are sparingly soluble in cold water and alcohol; soluble in boiling water and in acid and alkaline solutions. Warmed with strong nitric acid they become yellow; the residue, if touched with hydrochloric acid, becomes red, if with ammonia, brown. Millon's reagent (mercuric and mercurous nitrate) gives a red coloration resembling that given by proteid with the same reagent. Tyrosin, when it appears in the urine, is always associated with leucin, though this latter body may be present without tyrosin. It is found in small quantities in the spleen and pancreas, and is one of the products of the action of trypsin on albuminous matters.

60. **Hippuric acid** $C_9H_9NO_3$, or benzyl amido-acetic acid, containing radicals of benzoyl and glycocol, and may be written thus $C_7H_5O(C_2H_4)NO_2$. The crystals are semitransparent rhombic prisms; almost insoluble in cold water and ether, soluble in boiling water and solution of sodium phosphate. Boiled with strong hydrochloric acid it is decomposed into glycocin and benzoic acid. Hippuric acid is monobasic and gives buff-coloured precipitates with ferric salts. It occurs as a minute normal constituent of human urine, and in larger quantities among herbivorous animals.

61. **Lecithin** $C_{44}H_{90}NPO_9$ is an amorphous waxy substance, very hygroscopic; soluble in ether and alcohol, is precipitated from the solution by platinic chloride with excess of alcohol. Lecithin is decomposed by boiling with alkalies into glycerin-phosphoric acid, stearic acid, and cholin. *Protagon* has been considered as a mixture of lecithin and cerebrin, but the recent researches of Gamgee seem to show that it is a definite chemical body; he gives the formula $C_{160}H_{308}N_5PO_{35}$. *Cerebrin* is a nitrogenous body free from phosphorus. Some doubt still exists as to its composition. According to Gamgee, protagon, which cannot be separated by the action of solvents into a non-phosphorised cerebrin and a phosphorised body, can, however, by the action of caustic baryta be made to yield non-phosphorised bodies.

Group II. **Diamides.**

62. **Urea** CH_4N_2O, or carbamide, and may be represented as $\begin{rcases} CO'' \\ H \\ H \end{rcases} N_2$, in which the dibasic radical of carbonic acid has replaced two atoms of hydrogen; or as $CO \begin{cases} NH_2 \\ NH_2 \end{cases}$ in which 2 atoms of amidogen NH_2 have taken the place of 2 atoms of hydroxyl HO. Urea is also isomeric with ammonium carbamate and ammonium cyanide. Urea crystallises in colourless four-sided prisms which melt at 120° C. Very soluble in cold water, and its solutions are neutral to test paper. Heated to 150°C., urea is converted into bi-uret and cyanuric acid. With nitric acid urea forms nitrate of urea, which crystallises out in shining rhombic plates; these crystals are less soluble than urea crystals, therefore when the urine is concentrated they are often formed on the addition of nitric acid.

Oxalic acid also forms oxalate of urea, which is deposited as fine powdery crystals. Mercuric nitrate in alkaline solutions forms with urea an insoluble compound CON_2H_4, $4H_9O_1$. Hypobromous acid decomposes urea into water, carbonic acid, and nitrogen.

COMPOUND UREAS.

SUB-CLASS A.   MONUREIDES.

63. **Kreatin** $C_4H_9N_3O_2 + H_2O$. — This substance in contact with baryta water decomposes into urea and sarcosin (page 45). The crystals are oblique rhombic prisms slightly soluble in cold water and alcohol, very soluble in hot water, insoluble in ether. The solutions are neutral, and have an extremely bitter taste. Acted on by sulphuric acid, it is converted into kreatinin. Kreatin is chiefly found in the juice of flesh, and is undoubtedly an antecedent of urea.

Kreatinin $C_4H_7N_3O$ is formed from the foregoing by dehydration. It is an extremely powerful base, gives an alkaline reaction with test paper, and forms well-defined basic double salts with zinc chloride and silver nitrate.

SUB-CLASS B.   DIUREIDES.

64. **Uric acid** $C_5H_4N_4O_3$.— There is still some doubt as to the exact constitution of uric acid ; it is, however, represented by the hypothetical formula as consisting of one radical of tartronic acid and two of urea, thus :

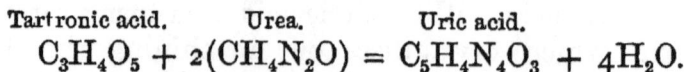

Tartronic acid.   Urea.   Uric acid.

$$C_3H_4O_5 + 2(CH_4N_2O) = C_5H_4N_4O_3 + 4H_2O.$$

From this it is held that uric acid is tartronyl cyanamide, four molecules of amidogen being replaced

by two of cyanogen, and two by the radical of tartronic
acid ; thus $N_2 \begin{cases} C_3H_2O_3 \\ CN_2 \\ H_2. \end{cases}$ As deposited from acid solutions it occurs in rhombic tablets of very variable form.
It is extremely insoluble in water and acid solutions,
very soluble in alkaline solutions. Uric acid is dibasic, and forms with bases both neutral and acid salts,
of which the sodium, potassium, and ammonium arc
of the most interest ; these are all very insoluble in
water, but less so than uric acid. When in solution,
if these salts are decomposed by the addition of
concentrated nitric acid, the uric acid is separated in
an amorphous and hydrated form, in which it is
rather more soluble than in its crystalline state. This
amorphous form of uric acid is frequently observed
when concentrated nitric acid is added to urine
when testing for albumin, and has been erroneously
described as precipitated acid urates, which may be
thrown down when dilute acid is used, if the urates
were previously in the neutral form. The characteristic test for uric acid is the *murexide* reaction, the
purple colour developed when the crystals are heated
with nitric acid, and the residue touched with ammonia. By oxydation uric acid yields alloxan and
urea ; mesoxalic acid and urea, allanturic acid and
urea; parabanic acid (which is oxalyl urea), tartronyl
urea or dialuric acid.

65. **Xanthin** $C_5H_4N_4O_2$.—The constitution of
this body is unknown ; uric acid treated with sodium
amalgam yields xanthin and hypoxanthin. Xanthin
forms white scales resembling beeswax, sometimes
deposited from urine in minute lemon-shaped plates ;
these dissolved in dilute hydrochloric acid, and the
solution slowly evaporated, yield hexagonal and prismatic crystals. Very insoluble in water (1 in 1,500),
freely soluble in dilute acids and alkalies. Heated

E

with nitric acid, and the residue moistened whilst still hot with liquor potassæ, a purple-red colour is developed. A constituent of a rare form of urinary calculus and gravel.

66. **Hypoxanthin** $C_5H_4N_4O$ (syn. *sarcine*).— Is a white imperfectly crystalline powder, rather more soluble in water than xanthin. Found in the spleen, thymus, muscular tissue, medulla of bones, and blood of leucæmic patients.

67. **Allantoin** $C_4H_6N_4O_3$ can be formed from uric acid by boiling with lead peroxide. Forms colourless hard glassy prisms of neutral reaction. Soluble in cold water (1 in 160). Boiled with potassium hydrate it yields potassium oxalate. Constituent of allantoic fluid and fœtal urine.

68. **Carnin** $C_7H_8N_4O_3$.—Has been discovered in Liebig's extract of meat; it can be converted into hypoxanthin by the action of bromine.

69. **Guanin** $C_5H_5N_5O$. — A yellowish-white powder nearly insoluble in water, but soluble in dilute acids and alkalies. By oxydation with potassium permanganate is converted into urea, oxalic acid, oxy-guanin. Is a normal constituent of the semi-solid excrement of birds. It has been sometimes met with in the liver, pancreas, and spleen, but does not appear to be a constant product.

DIVISION III.—VEGETO-ALKALOIDS.

These bodies are supposed to be compound ammonias, and act as powerful bases ; are the active principles of certain plants. Those of highly poisonous nature are enumerated here, as they may be introduced into the animal body either accidentally or with criminal intent, when their detection becomes a matter of consequence. A group of animal alkaloids, the result of putrefactive changes in the tissues, have recently been discovered, and are named *ptoamines ;* they correspond in their general reactions with the vegeto-alkaloids.

70. **Morphine** $C_{17}H_{19}NO_3 + H_2O$.—The crystals are colourless prisms, very slightly soluble in water (1 in 1,000 cold; 1 in 500 hot). Very soluble in hot alcohol. The salts are more soluble in water than the base. It gives an indigo-blue coloration, with a neutral solution of ferric chloride. With strong nitric acid forms a deep orange-yellow compound. Concentrated sulphuric acid, with a trace of nitric acid, gives a violet-purple colour. Morphine decomposes iodic acid with the liberation of iodine.

71. **Strychnine** $C_{21}H_{22}N_2O_2$.—The crystals are extremely small brilliant octohedra, transparent, and colourless. Slightly soluble in water (1 in 6,000 cold; 1 in 2,500 hot), more soluble in chloroform. The solutions have an intensely bitter taste (perceptible 1 in 1,000,000). With concentrated sulphuric acid, and a fragment of potassium dichromate, a deep violet tint is produced, gradually fading on exposure. Solutions of strychnine ($\frac{1}{1500}$ grain) injected under skin often produce violent tetanic spasms and death.

72. **Brucine** $C_{23}H_{26}N_2O_4 + 4H_2O$.—More soluble in water than strychnine, and is readily soluble in alcohol. Is distinguished from it by the bright red colour given when touched with nitric acid.

73. **Curarine** $C_{10}H_{15}N$.—Very soluble in water and alcohol. Nitric acid gives a purple coloration. Concentrated sulphuric acid colours it blue, and if potassium dichromate is added, changes to violet slowly fading to yellow. In this test it resembles strychnine, but is distinguished from that body by its ready solubility in water.

74. **Atropine** $C_{17}H_{23}NO_3$.—Crystallises in colourless needles, slightly soluble in water, very soluble in chloroform. With concentrated sulphuric acid

no coloration in the cold, but on heating a yellow
tinge is developed, and on adding water a rose-
like odour is given off.   Solutions containing trace of
atropine have highly poisonous effects on frogs.

75. **Ptoamines.**—This name has been given to
bodies which have been detected in exhumed corpses,
closely resembling, in their chemical reactions and
physiological effects, the vegetable alkaloids.   Whilst
some, however, act as powerful poisons, others are
inactive, and a few actually counteract the effect of
poisonous substances.   Spica, in the liquid from a
suppurative peritonitis, obtained bodies, some of
which were oily and volatile, with a strong alkaline
reaction and capable of forming crystalline salts, and
having the odour of conine.   The chloroform extract
was extremely poisonous in its actions on frogs, and in
its physiological effect resembled that of curarine.
Sonnenschein has found, in an anatomical maceration
fluid, an alkaloid resembling atropine in its action.
Ranke has shown that the physiological action of
strychnine in bodies long buried may be masked by
ptoamines.   A body resembling Sonnenschein's alkaloid has been found in bodies of patients dying
from typhus fever.   It is interesting to observe that
in many instances of death resulting from poisonous
food, the patient showed marked typhus symptoms.
It appears that these bodies are generally produced in
substances which, after brief exposure to air, are
buried or excluded from air, as in corpses, tinned
meat, etc., and that then they are found in the most
internal portion.   It is of importance to discover the
reactions which distinguish these bodies from the
vegeto-alkaloids, but the question is often one of
extreme difficulty.   It has been stated that potassium
ferrocyanide is a reliable reagent, since ptoamines
reduce it, whilst the vegeto-alkaloids have no reaction
on it.   The test is applied as follows :—The extract

is converted into a sulphate, and a few drops of solution of K_4FeCy_6 added; if a ptoamine is present, prussian blue will be formed on the addition of a few drops of ferric chloride. No reaction, however, is given with a vegeto-alkaloid, except morphine and veratrine; these, however, can be detected by their special tests.

INDIGO GROUP.

76. **Indol** C_8H_7N.—A crystalline substance which is the starting point of the indigo group. It occurs in the body as one of the products of pancreatic digestion, and is said to give the characteristic odour to the fæces. The addition of salicylic acid to pancreatic juice stops its formation from albuminoids; and it has been found that under the administration of the acid the quantity of indol decreases exactly in proportion as the quantity of phenol increases. Very dilute nitrous acid gives a red colour in solution of indol.

Indigo $C_{16}H_{10}N_2O_2$.—This substance, which is derived commercially from the *indigofera*, is sometimes met with in sweat and urine, giving rise to a bluish colour. It has been met with as a constituent of urinary calculus (*Med. Path. Soc. Trans.*, vol. xxix., p. 157). The source of indigo in the body is undoubtedly from indol formed by the decomposition of albuminoids by pancreatic digestion, since *indican* $C_{52}H_{62}N_2O_{34}$, or glucoside of indigo, is found in the urine of animals after the subcutaneous injections of indol, also after ligature of the small intestine, and also, according to Senator, in obstructions and other affections of the intestines in disease. The indol being probably converted into indican in the alkaline blood, but deposited as indigo when it comes in contact with the acid urine. However this may be, indican out of the body always yields indigo as one product of its

decomposition, and *indi-glucin*, a sweet, non-ferment-
able substance which reduces Fehling's solution.   Uro-
xanthin, a colouring matter found in the urine, strongly
resembles indican ; it does not, however, yield indigo-
blue under the action of acids.

78. **Colouring matters of human body**
are the colouring matters of the blood, hæmoglobin
and its derivatives.   The bile pigments, bilirubin,
biliverdin, bilifuscin, biliprasin, urobilin.   The urinary
pigments, uroxanthin or indican, and urobilin and a
black pigment melanin, found normally in the choroid
and in the skin of negro, and pathologically in melanotic
tumours, in the urine, and as a deposit in the lungs.
For the reaction of the various pigments, the reader
is referred to chapters on blood, bile, and urine.

SECTION B.—INORGANIC CONSTITUENTS.

79. **Hydrochloric acid** (HCl). — Chlorine in
the body is chiefly found in combination with the alkaline oxides of soda and potash.   These salts are, however,
variously distributed; thus, for example, in 1000 parts
blood corpuscles, we find 3·67 parts of potassium with
only a trace of sodium chloride, whilst in the plasma
there is only ·36 parts of potassium chloride, and as
much as 5·54 parts in the 1000.   The chlorides appear
to fulfil the following purposes in the economy : (1)
by furnishing the hydrochloric acid for digestion
($2Na_2HPO_4 + 3CaCl = Ca_32PO_4 + 4NaCl + 2HCl$;
Maly) ; (2) by aiding the metabolic processes going
on in the body, an increased ingestion of common salt
being followed by increased excretion of urea.   Sodium
chloride is found in great abundance in all cellular
growths, as in cartilage, mucus, etc.   In certain
diseases attended with increased cell formation, as in
cancer, in pneumonia with exudation, and in purulent discharges, sodium chloride is present in large

quantities in the morbid products, and consequently there is a sensible diminution in the quantity excreted by the urine. *Silver nitrate* gives, in extremely dilute solutions, a faint haze; in stronger solutions a white cloudy precipitate; nitric acid will not cause these to dissolve; they are soluble, however, in ammonia, but on nitric acid being added to the ammoniacal solution, the silver chloride is again precipitated.

80. **Hydrofluoric acid** HF. — United with calcium as calcium fluoride, is found in minute quantities in the bones and teeth, occasionally in blood, milk, and urine. The ash treated with strong sulphuric acid, and gently heated in a glass vessel, the glass becomes corroded. Large quantities of ash must be employed.

81. **Phosphoric acid** H_3PO_4.—Phosphorus is one of the most important of the inorganic constituents of the body, and is widely distributed. In combination it is found as a component of the complex nitrogenous fats, lecithin, etc. In its oxydised form it occurs as tribasic phosphoric acid. In combination with soda and potash it forms the alkaline phosphates, the neutral salts of which (Na_2HPO_4), together with the alkaline carbonates, gives to blood its alkaline reaction; the acid salts NaH_2PO_4 gives urine its acid reaction. Bone earth is tricalcic phosphate Ca_32PO_4. Magnesium phosphate associated with calcium phosphate is found in all animal tissues and fluids, but to a less extent; in the excreta, however, the magnesium is relatively more abundant than the calcium salt. From this it has been inferred that the magnesium salt is less required by the organism, and consequently is not so long retained as the calcium salt, and also that less is absorbed by the intestinal canal. Few questions in scientific medicine afford less facts for generalisation than the variations in the excretions of phosphoric acid by the urine in disease. Physiology can only tell us that the element phosphorus is

absolutely essential for the growth and nutrition of the tissues, but cannot explain its rôle. Pathologically we find (a) excess of phosphoric acid accompanied by a proportionate increase in the elimination of urea ; (b) excess of phosphoric acid without a proportionate increase. In the first instance the increase is probably due to increased tissue metabolism generally ; in the second, probably only the nervous system is involved. The compound known as *triple phosphate,* or ammonio-magnesium phosphate $PO_4Mg(NH_4) + 6H_2O$, is met with in ammoniacal urines ; this salt combined with calcium phosphate forms the basis of one of the most frequent forms of phosphatic calculus, known as the fusible calculus. Phosphoric acid gives a yellow precipitate with *silver nitrate* soluble in both nitric acid and ammonia. A yellow precipitate is given with *ammonium molybdate* in nitric acid. An acid solution of *uranic nitrate* is decomposed by phosphoric acid, and uranic phosphate is precipitated ; this latter gives no stain with ferrocyanide of potassium test-paper, which uranic nitrate does ; upon this fact the process for estimating quantitatively phosphoric acid is based. (*Q.v.* Urine.)

82. **Sulphuric acid** H_2SO_4.—Sulphur is found in most protein bodies, and in many products of their metabolism, as cystin and taurin. It is also largely introduced into the body with vegetable substances. Only a small portion of the sulphur in the body passes out as sulphuric acid, a considerable part being discharged either unoxydised or only partially oxydised by the bowels, and some by the skin and epidermal appendages. It is thus difficult, from the increase or diminution of sulphuric acid in urine, to form an estimate of the degree of metabolism of the sulphur compounds in the body. The sulphuric acid in the urine is for the most part combined with potash and to some extent with lime. About 0·4 grain of

sulphur also passes into urine in a partially oxydised state in the twenty-four hours. The sulphates are detected in acid solutions by the addition of *barium chloride* or *nitrate*, which throws down a dense cloud of barium sulphate.

83. **Sulpho-cyanate of potassium** CNKS. —A trace of this substance is found in saliva. Pettenkoffer believes it to be derived from decomposition occurring in the cavity of the mouth, being derived from urea and sulphate of potash. Dr. S. Fenwick, however, has (*Med.-Chir. Trans.*, 1882) endeavoured to show that the presence of sulpho-cyanate in saliva is not due to accidental causes, but is regulated by the operation of general laws, and the amount present varies considerably in disease. The chief point of practical importance is its relation to medico-legal inquiries in cases of opium poisoning, since meconic acid gives the same reaction with *ferric chloride* as the sulpho-cyanate, viz., a cherry-red ; this, however, disappears on the addition of mercuric chloride, which is not the case with the colour produced by meconic acid.

84. **Lime** CaO. — Lime must be considered as one of the chief mineral constituents of the body. As phosphate of lime, it ranks next to water in the importance of its physical properties to the organism. Its salts constitute 64 per cent. of the weight of the skeleton, viz. : calcium phosphate, 51·04 ; calcium fluoride, 1·9 ; and calcium carbonate, 11·4 per cent. In blood the calcium salts are in excess in the plasma, 1000 parts yielding ·298 part of lime, whilst the same weight of corpuscles give only ·094. In muscular tissue lime is present, ranging from ·230 to ·311 part in a 1000. Lime salts are tolerably abundant in the gastric juice, about 1·48 of calcium phosphate in the 1000. Maly believes the hydrochloric acid of the gastric juice is derived from the decomposition of chloride of calcium($2Na_2HPO_4 + 3CaCl = Ca_32PO_4 + 4NaCl + 2HCl$).

Only a small proportion of the lime injected is excreted with the urine. In experiments on dogs it was found that when calcium chloride was given by the mouth nearly all the lime was passed off by the bowel as carbonate, only a small increase in the excretion of lime by the urine being noted, and that as phosphate. The chlorine appeared in combination with sodium as chloride of sodium. In rickets it has been said there is a great increase in the quantity of lime salts excreted in the urine, but the fact is by no means established; recent observations have actually found a diminution, which was most marked when the disease was at its height. It is probable that the idea that an increase of lime excreted took place in rickets was due to the mere deposition of the earthy salts being taken for excess. The urine of rickety children is frequently alkaline, and consequently is often turbid from deposited calcium phosphate. Local deposits of calcium carbonate are frequently met with in tissues which have suffered from general impairment of vitality, or from diminished nutritive supply. In these cases it is probable that, owing to the escape of the free carbonic acid, which keeps this salt in solution, from the stagnated fluids, the calcium salt is precipitated, and on account of the degeneration of the tissues is not taken up by them. Salts of lime form the insoluble portion of the ash; to obtain them in solution they must be dissolved in as small a quantity as possible of dilute acetic acid. *Ammonium oxalate* added to this solution gives a copious white precipitate of calcium oxalate. Mineral acids should not be used to dissolve the ash, as the calcium oxalate would redissolve if the acid was in excess. The amount of lime present is best determined by weight. For this purpose the insoluble portion of the ash is weighed, and the weight recorded. It is then dissolved with acetic acid, and the solution diluted by the addition of about ten times its weight of distilled water. To this solution add saturated solution of ammonium oxalate, and allow to stand for twelve hours. Filter through a filter the weight of whose ash is known.* The precipitate in the filter consists of calcium oxalate. The filtered solution is magnesia, which should be set aside for determination of that substance if required. Wash the precipitate on the filter with distilled water. Dry over water-bath till the filter and contents cease to lose weight. Place the filter and contents in a small platinum capsule with cover, whose weight is known; heat till whole is reduced to white ash. When cool convert the lime into sulphate by the addition of a few drops of sulphuric acid. Again heat carefully, covering to prevent loss. Weigh the capsule when cold. Now, when the weight of the ash of the filter and the weight of the capsule are deducted from the total weight, we have the weight of sulphate of lime, and to calculate this as caustic lime, we multiply by 0·4118.

85. **Magnesia** MgO. — Magnesium phosphate associated with calcium phosphate is found in all the animal tissues and fluids, but in smaller quantities. Little is known as to the purpose it fulfils in the economy. The chief clinical interest attaching to it is from it forming, as triple phosphate, a variety of urinary gravel, and a calculous crust in certain morbid conditions of the urinary organs, which induce the formation of volatile (ammonia) alkali. The triple phosphate is, however, generally associated with calcium phosphate, the mixture of the two salts forming a mass readily *fusible* under the blowpipe. Neutral solution of magnesia throws down a bulky gelatinous precipitate with *ammonia*. The addition of *sodium phosphate* to a magnesian solution

---

* This is done by incinerating a filter of the same size and weight as the one used, and weighing the ash.

gives rise to a white precipitate of magnesian phosphate, whilst the addition of *ammonia* throws the
ammonio-magnesium phosphate.   Magnesia is estimated by weight.   The ash is treated, in the first
instance, as directed for lime.   The filtrate, obtained
after the removal of the precipitate caused by ammonium oxalate (lime oxalate), is treated with excess
of ammonia, and allowed to stand twelve hours.
The resulting precipitate is collected on a filter, the
weight of whose ash has been previously ascertained ; the filter is then dried till it ceases to lose
weight.   When dry it is to be placed in a small
platinum capsule whose weight is known, and heated
to a white heat till a white glassy mass is left at the
bottom of the capsule.   This is magnesium pyrophosphate.   Weigh the capsule when cold, and by
deducting the weight of the ash of the filter and
the weight of the capsule from the total weight, the
result gives amount of magnesium pyrophosphate ;
by multiplying this weight by 0·3604 we get the
amount of magnesia present as oxide.

86. **Ammonia** NH_3. — Ammonia in a free
state is sometimes found in urine ; generally, however,
it is in the form of carbonate, the result of ureal decomposition.   It has been detected in the breath of
typhus patients.   The chief salt of interest it forms
is ammonium urate, a constituent of the urinary
excretion of insects, reptiles, and birds, and is frequently met with as a urinary deposit in human
beings.

87. **Potash** KHO. — This body, with soda,
forms the soluble portion of the ash.   Potash is
widely distributed throughout the body, but in very
variable proportions.   It is more abundant in the
solid tissues and the corpuscles than in the secretions and plasma of the blood ;   in this respect
being the very opposite of the soda salts.   The

potash salts seem to have a greater activity than
those of sodium. Ringer has found them more
poisonous, as far as their action on the heart is
concerned, than the sodium salts. It is said that
a dog fed on nothing but Liebig's extract dies
sooner than a dog not fed at all, on account of
the potash salts exerting a poisonous influence. Dr.
Garrod, some years ago, found that the amount of
potash salts in urine were diminished in scurvy,
and suggested that this disease was in some way
connected with a deficiency of this base in the
body, and that dietaries of persons affected with the
disease were deficient in potash. Further examination
has not shown this to be correct, since peas and other
articles of a sailor's diet have been shown to contain
a sufficiency of potash. I have, however, recently
confirmed Dr. Garrod's observation as to the deficiency of potash in the urine of scorbutic patients,
but I believe that the diminution is not due to
diminished ingestion, but to the base being withheld
by the blood in order to maintain its alkalinity,
which has been impaired by the withdrawal of the
alkaline carbonates supplied by fresh meat, vegetables, and fruits. Solutions of potash salts give,
with platinum bi-chloride, when slightly acidulated
with hydrochloric acid, a yellow crystalline precipitate of potassium platinic chloride. The estimation
of potassium and sodium can be conveniently done
together, and will be considered in the next paragraph.

88. **Sodium** NaO is more abundantly met with
in fluids than in the solid textures of the body, in this
case being the reverse of the potassium salts. Thus,
we find in blood plasma 5·546 parts of sodium chloride,
1·532 parts of soda, and ·271 part of sodium phosphate,
as compared with ·359 part of potassium chloride and
·281 of potassium sulphate in 1000. Whilst in the

blood corpuscles the reverse obtains ; for in 1000 parts we have 3·679 potassium chloride, 2·343 potassium phosphate, ·132 potassium sulphate, against ·341 of soda and ·635 sodium phosphate. The acidity of the urine is chiefly due to acid sodium phosphate NaH_2PO_4. The bile, too, is particularly rich in soda salts, containing about 2·5 per cent. of sodium chloride and 6 per cent. bile salts, in the form of glycocholate and taurocholate of soda. The action of soda salts on the excitability and contractility of the heart is decidedly less than that of potash salts, but it is nevertheless marked. The chief clinical interest attaching to salts of soda is the formation of gouty tophi and calculous infarcts in the tubules of the kidney, composed of sodium urate. A concentrated solution of sodium salt gives a white crystalline precipitate with *antimoniate of potash*. Soda salts also give a characteristic yellow tinge to the blow-pipe flame. Potassium and sodium can be determined quantitatively by the same operation. Dissolve a definite quantity of ash in boiling distilled water. Filter. Bring the filtrate up to 60 cc. by the addition of distilled water, or diminish by evaporation to that bulk. Of this quantity take 30 cc. (which will represent half the quantity of the soluble salts in the ash), add a few drops of ammonia and ammonium carbonate. Set aside for twelve hours, filter through a very small filter. Acidulate the filtrate with a few drops of hydrochloric acid, and place it in a platinum capsule of known weight. Evaporate to dryness, and then heat gradually to redness to drive off ammoniacal salts. Weigh when cool ; deduct weight of capsule ; result *gives weight of potassium and sodium chlorides*, of half the amount of ash originally taken for experiment. To estimate the potash, dissolve the residue in the platinum capsule in as little water as possible, and then add an alcoholic solution of

platinic bichloride till the solution acquires a deep
yellow colour; evaporate to near dryness, and then
add 50 cc. of absolute alcohol and 10 of ether.
Set aside for twenty-four hours, frequently stirring.
Collect precipitate on weighed filter; wash it with
alcohol; dry it. Weigh; deduct weight of filter.
The remainder *gives amount of potassium platino-
chloride* (100 parts of which equals 30·51 potassium
chloride). Now, as we have learnt the combined
weight of the potassium and sodium chlorides from
the first part of the process, we have only to
deduct the weight of the now ascertained potassium
chloride from the weight of the two chlorides, to find
the difference, *which is the amount of the sodium
chloride.* To get the amount of potash and soda we
have to multiply the weight of potassium chloride by
0·6317 and the sodium chloride by 0·5302. The
result *being the amount of potash and soda* in the
ash, in the 30 cc. of the dissolved ash, or exactly half
the amount of the whole ash dissolved in the first
instance. The object of dividing the original solution being, of course, to reserve a portion in case of
accidents, or a desire to check the process at any
point

89. **Iron** F_2O_3.—Traces of iron are met with
in the ash of most tissues and fluids; as a proto-
chloride in the ash of gastric juice, and as a phosphate
in muscular and splenic juice. Combined with
globulin as hæmoglobulin, it forms the colouring
matter of the blood. Preyer from an average of eleven
cases gives 0·056 grm. of iron and 13·45 grms. of
hæmoglobin in 100 grms. of human blood. It is
probable that the iron of the effete blood-corpuscles
passes away with the bile, recent observations
tending to show that the colouring matters of the bile
are produced from hæmatin by reduction, due to the
action of the bile acids on hæmoglobin. When iron

salts are taken an increase of this metal is found in the bile. The salts of iron are met with as proto or ferrous salts, and per or ferric salts. The former give white precipitates with caustic alkalies and light blue precipitates with potassium ferrocyanide ; the latter a reddish-brown precipitate with caustic alkalies and a deep blue precipitate with potassium ferrocyanide. The amount of iron in the ash of any tissue or fluid is determined by the standard solution of potassium permanganate. This was formerly the only method by which the amount of iron in blood could be determined. Since, however, the proportion of iron to hæmoglobin has been ascertained, the metal can be estimated quantitatively through the determination of the hæmoglobin, either by Hoppe Seyler's or Gower's method. The procedure by these methods will be treated more conveniently when we consider the variation occurring in the amount of hæmoglobin in disease. (*Vide* Blood.)

90. **Silica** S_1O_2.—*Silica* is a constituent of the epidermal tissues ; it is nearly always present in the fæces, and occasionally in the blood, bile, and urine. Silica may be obtained from the ash of any of the substances in which it is present, by fusing the ash with eight times its weight of sodium carbonate, and boiling the mass in water ; on the addition of hydrochloric acid the silica is partially precipitated as a gelatinous mass. The acid solution is now evaporated, and the residue treated with some more hydrochloric acid and dried. The silica will then be left as a white insoluble powder.

91. **Lead** Pb; **arsenic** As_2O_5; **copper** Cu. —These substances are only incidentally present in the tissues, and are apparently in no way necessary to the maintenance of their functions. It has been stated that arsenic and copper are constantly present in minute quantities in the human body. The method

of separating lead and arsenic from the tissues and
fluids in cases of poisoning is detailed in the chapter
on morbid digestion and on urine.

CHAPTER III.

BLOOD—CHYLE—LYMPH—MILK.

Examination of morbid blood.—Of all
branches of pathological chemistry, less advance has,
perhaps, been made in determining the changes
which blood undergoes in disease than in any other
direction. Some excuse may be offered for this
apparent neglect, in the difficulty of obtaining blood
in sufficient quantity for analysis, a difficulty that did
not exist in days when bleeding was general and
regarded as part of the ordinary treatment. Also, the
methods of research until recently have been very
defective. The study of the chemistry of morbid blood,
however, must be seriously undertaken before we can
expect to judge of the effects produced by even
primary alteration of the blood in disease. For as yet
how little do we know of the variations that daily and
hourly occur in the chemical composition of normal
blood, and what is the action, physical as well as
chemical, that each of its constituent elements has on
the albumins, fats, salts, and water that compose the
tissues, and how far excess or diminution of these
constituents influences oxydation and nutrition in
the body.

92. **Variation between the water and
solids.—Specific gravity.**—The following figures
give approximatively the composition of normal blood
in 100 parts. Water, 79·5; solids, 20·5; serum

F

albumin, 7·2; fibrin, 0·21; hæmoglobin, 11·5; fatty
matters, 0·18; extractives, 0·32; ash, 0·81. 100 parts
of blood serum contain water 90·5 parts; solids,
9·6. The proteids varying from 8 to 9, and the
fat extractives and salts from 2 to 1. 100 parts
of wet blood-corpuscles yield water, 56·5; solids, 43·5;
hæmoglobin, 41·1; other proteids, 3·9; fats, chiefly
cholesterin and lecithin, ·37. The proportion of water,
however, varies considerably, and probably ranges from
76 to 80 in healthy blood under the influence of
normal physiological conditions, the ingestion of fluids,
of solid food, and the activity of the renal and
cutaneous excretions. The proportion of water to the
solids can be ascertained by evaporating a small
quantity of blood in a weighed platinum capsule, till
it ceases to lose weight. Then, if z represents the
weight of the blood evaporated and y the loss sus-
tained by evaporation, then $\dfrac{y \times 1000}{z} = x$ the proportion of water in 1000 parts. The specific gravity
is ascertained by means of a small flask fitted with
perforated stopper for the insertion of a thermometer.
The flask is then filled with distilled water at 15° C.,
and weighed; it is then emptied and dried, and
then filled with defibrinated blood at the same
temperature and again weighed. Then if a represents
the weight of the distilled water and b the weight
of the defibrinated blood, $\dfrac{a}{b}$ = the specific gravity of
the blood; thus, if the weight of the flask filled with
distilled water is 25·573 grains, and the weight of the
flask with defibrinated blood is 27·160 grains, then
$\dfrac{27 \cdot 160}{25 \cdot 573} = 1 \cdot 060$. As, however, liquids expand and
contract by alteration of temperature, a correction is
necessary if the two fluids are weighed at different
temperatures. Practically, however, if both weighings

are conducted at the same time and in the same room
very little difference occurs; if it does, it can be
corrected by plunging the bottle in warm water if the
temperature be too low, or into cold water if too high.
The following gives the maximum and minimum
percentage amount of water in different diseases as
recorded by trustworthy observers. Normal blood,
78·8—80; scurvy, 84·9—83·5; chlorosis, 86·8—81·8;
cancer of liver, 88·7; chronic Bright's disease, 80·8—
88·7; puerperal eclampsia, 77·8—80; diabetes mellitus,
79·4—80·2; cholera, 74·0—75·0. There are many
analyses of blood in acute diseases, but their value is
impaired by the fact that they were generally made
after repeated venesections, which of course in itself
renders the blood more watery.

93 **Reaction.**—Acid and acid salts are continuously entering the blood. (1) They may be introduced
into the body from without in the food. The quantity,
however, thus derived under ordinary conditions is
comparatively small, since nearly the whole of the
saline constituents of the food are alkaline, or become
so by conversion in the system. Still, a small quantity
of acid sodium phosphate is derived from the juice of
flesh, and this passes no doubt unchanged into the
blood. (2) Acid, too, is formed in the alimentary
canal from fermentative decomposition of the saccharine
matters taken with the food, or of the amylaceous
principles that have been converted into sugar. (3)
Lastly, acid is generated in the tissues of the body.

In spite, however, of this constant entrance of acid
into it, the blood of the living body is always alkaline,
no doubt because the chief acid salt (sodium bicarbonate) has an alkaline reaction. What the degree of
alkalescence of normal blood is has not been determined,
but it is probable that it has certain definite limits
which cannot be passed in either direction without
causing disturbance of healthy nutrition. In fact, great

difficulty is experienced in reducing the alkaline reaction of the blood. Hoffmann, who fed pigeons for a considerable time on food yielding only an acid ash (yolk of egg), found that however great the tendency of uric acid and of the acid salts of phosphoric acid to combine with bases, yet these were not withdrawn from the alkaline blood, but were evidently withheld to maintain its alkalinity. Loscar, by introducing diluted mineral acids into the stomach, succeeded in reducing the alkalescence of the blood, but not considerably, and the conclusion he arrived at was that the organism retained free alkali with great energy. In some of his experiments the quantity of acid introduced into the stomach would have made the whole animal acid if it had been absorbed and excreted again. From this Loscar infers that the organism possesses certain " regulative mechanisms " which maintain the equilibrium between the acids and alkaline bases in the system. These experimental facts seem to be borne out by what occurs in scurvy. That disease, as has been well established, is brought about by the prolonged and complete withdrawal of the organic vegetable acids and their salts from the dietary of those affected. These organic salts, as is well known, by oxydation in the blood yield alkaline carbonates. Now, the alkaline carbonates are the salts chiefly concerned in maintaining the alkalescence of the blood, and it appears when these are largely withdrawn, as happens when scurvy is induced, the proper degree of alkalescence of the blood is maintained with difficulty, and in order to secure it some other alkaline salt is retained instead of being excreted.* Thus, I found, after the withdrawal of fresh vegetable food for eighteen days, the total quantity of phosphoric acid passed in the twenty-four hours was slightly reduced, whilst the phosphoric acid in combination with the

* "General Pathology of Scurvy,." Lewis. 1877.

alkaline oxides was reduced nearly one half. Again, in a case of scurvy, it was found that the alkaline phosphates increased rapidly on the resumption of an antiscorbutic diet, although the amount of phosphoric acid ingested was the same in scorbutic and antiscorbutic rations respectively; these two facts pointing to the conclusion that the alkaline phosphates are retained in the system when the alkaline carbonates are withdrawn, and discharged when these are again supplied. All experiments made on animals with a view to reduce the alkalinity of the blood or to neutralise it have ended sooner or later in the death of the animal; and if the process has been a slow one, the definite pathological changes will be found to have occurred in the blood and tissues, closely resembling the changes found in the bodies of patients dying from scurvy, viz. dissolution of the blood globules, ecchymosis in the heart, blood stains in the mediastinum, gums, and mucous surfaces; whilst the muscular structure of the heart, and the muscles generally, as well as the secreting cells of the liver and kidneys, become granular and even distinctly fatty.   Lastly, Dr. Gaskell has shown, experimentally, that a dilute alkaline solution acts upon the muscular tissues of the heart so as to produce a powerful contraction, whilst dilute acid solutions produce an opposite effect, and that the muscles of the smaller arteries are acted upon in the same way. These facts seem to point to the conclusion that one factor at least upon which the constriction of the muscles both of heart and arteries depend, is the alkalinity of the fluid surrounding them.   It is not unreasonable, therefore, to surmise that variations of the degree of alkalinity would not be unlikely to lead to disturbances of circulation and so effect a secondary chemical influence on nutrition, as well as a direct one.

The two salts chiefly concerned in the maintenance of the alkalinity of blood are neutral sodium phosphate

Na$_2$HPO$_4$, and acid sodium carbonate (bicarbonate) NaH$_2$CO$_3$; it is owing to this composition that the seeming paradox of the separation of acid secretion, such as urine and the gastric juice, from the alkaline blood can be explained. (*Vide* §§ 8, 107, and 134.) No precise observations have been made as to variations in the alkalinity of the blood in disease. Garrod has found it diminished in a case of gout; it has also been found lessened in the blood serum in chronic Bright's disease and in cholera. This particular branch of enquiry ought to yield abundant information to the clinical enquirer, with respect to the nature of such diseases as acute rheumatism, gout, scurvy, and diabetes mellitus. The serum from a blister may be used, or sufficient blood for the purpose can be readily obtained by means of the artificial leech. This should be defibrinated, and placed in a beaker. Then from a Mohr's burette, a standard solution of crystallised tartaric acid, 1 cc. of which is equivalent to ·004 grm. sodium hydrate, is to be added. This solution is gradually dropped into the beaker containing the blood, and a drop transferred after each addition by means of a glass rod to a plaster of Paris plate, stained with a solution of blue litmus. The addition of the acid solution is continued till a faint red stain is given to the plate by the drop placed in it. The number of cubic centimetres of the standard solution employed to neutralise the blood, multiplied by ·004, gives the alkalinity of the quantity of blood taken in terms of sodium hydrate. In preparing the plate care must be taken to obtain the plaster of Paris free from alkaline reaction, and to remove any alkali from the litmus. This is done by thoroughly dissolving 20 grammes of litmus in 150 cubic centimetres of water, and allowing the solution to stand twenty-four hours, then filtering. The precipitate is again dissolved in 250 centimetres of distilled water, and allowed to stand for twenty-four

hours; it is then divided into two equal portions.
One is treated with dilute acid, added drop by
drop by means of a glass rod till the solution is
faintly red; it is then added to the other portion.
The plaster of Paris plates are then soaked in
the resulting violet-blue solution. It is absolutely
necessary to prepare the plate before obtaining the
blood, since by keeping, even a few hours, the
alkalinity of the blood is diminished.

94. **Fibrin.**— When blood is drawn from the
body it separates into clot and serum. The former
consists of fibrin, holding in its meshes the blood
corpuscles; these latter can be removed by washing
the clot with a gentle stream of water. The amount of
fibrin in normal blood, as according to recent analyses,
ranges from ·21 to ·23 parts in 100. The older
analyses state it as ·31 to ·34. In many diseases it is
considerably increased. The following gives some of
the more reliable observations : Acute bronchitis, from
·43 to ·63 in 100 parts. Pneumonia, ·40 to 1·05.
Pleurisy, before effusion, ·58 to ·59 ; pleurisy, after
effusion, ·40 to ·48. Acute rheumatism, the mean
of forty-three observations, ·67. Puerperal convulsions, ·44 to ·60. Heart disease, mean of twenty-four
cases, ·34 ; advanced heart disease, ·25, mean of thirty-
one cases. Chlorosis, mean of nine cases, ·29. Scurvy,
·45 to ·65. Diabetes, ·19 to ·24. Phthisis, mean of
twenty-two analyses, ·44. Puerperal fever, ·43 to ·51.
The so-called " buffy coat," which sometimes forms
on the surface of the clot in blood drawn in certain
diseases, does not necessarily depend on the amount
of fibrin present, but on delay in the act of coagulation, and the tendency of the red corpuscles to
aggregate together, which is generally observable
in the inflammatory state. These, therefore, have
a tendency rapidly to subside, leaving the white
corpuscles, which are lighter in the upper portion

of the clot, and these give it the buffy appearance. The antecedents of fibrin in the blood are paraglobulin, or *fibrino-plastic* substance, and *fibrinogen;* whilst for the conversion of fibrinogen into fibrin outside the body, the action of a third body, the *fibrin ferment*, is required (§ 21). The addition of this latter body to fluids containing fibrino-plastic and fibrinogen immediately brings about coagulation, as may be seen sometimes in its addition to chylous urine, in which there is but little coagulum, causing at once an abundant separation. This is also occasionally observed with hydrocele fluid. Fibrin is estimated quantitatively in uncoagulated blood by drawing the blood directly into a bottle, whose weight has been accurately determined, fitted with a perforated stopper, through which a rod is inserted, to which is attached a brush of fine wire. When a sufficiency of blood has passed into the bottle, the rod with attached whisk is stirred round and round for a quarter of an hour, when the fibrin will attach itself to the wires composing the brush. The outside of the bottle is now to be carefully cleaned and dried, and then weighed; the addition to the original weight of the bottle and apparatus gives the amount of blood employed. Now pour the contents of the bottle into a muslin strainer, and with a pair of delicate forceps pick off from the wires the adhering fragments of fibrin, and place them on the muslin strainer, wash most thoroughly with cold water till they are completely freed from adhering corpuscles or serum, and subsequently wash with ether. Then remove them to a weighed platinum capsule, and dry over water-bath till it ceases to lose weight, and let it cool in chamber of balance, over a beaker of strong sulphuric acid. Then weigh; the increase of weight in the platinum capsule gives the amount of fibrin. This, however, contains some inorganic residue; in order to obtain the fibrin pure it must be incinerated. For this purpose the capsule is heated till only a white ash is left, and is again weighed. The loss in weight deducted from the weight of the fibrin before incineration gives the amount of pure fibrin. Supposing the original weight of blood was 31·2 grammes, and the weight of pure fibrin obtained from this ·066 grammes, then

$$\frac{100 \times ·066}{32·2} = ·21$$ grammes of fibrin in 100 parts of

blood. When, however, we have to deal with blood already coagulated we have to cleanse the already separated fibrin from the blood corpuscles, a proceeding of considerable difficulty. This is done by weighing 100 centimetres of blood in a beaker, and draining off the serum through a linen strainer, washing well with cold water; after a while fold the strainer, as a bag, over the contents; wash with water and knead the contents till all the corpuscles are removed, and the clot decolorised. Then transfer the filaments by means of a delicate pair of forceps to the platinum capsule, and proceed as above.

95. **Colouring matter of the blood** is contained in the red corpuscles. These are minute discoidal bodies, nearly transparent, of a yellowish colour, varying in diameter from $\frac{1}{3000}$ to $\frac{1}{4000}$ of an inch, the average being put at $\frac{1}{3200}$ (or 7·9 micromillimetre), and about $\frac{1}{12500}$ inch (1·8 micromillimetre) in thickness. The human blood corpuscles contain no nucleus. In fresh unaltered blood the surfaces of the red corpuscles are bi-concave, and form peculiar rolls, like coins adhering together. Under certain abnormal conditions, or when the blood is diluted with saline solutions, as sodium chloride, sodium sulphate, magnesium sulphate, etc., the corpuscles lose their smooth circular outline, shrinking, and becoming crenate. On the other hand, when placed in fluids of low density they swell up, and become

bi-convex and globular, and may even burst. The red corpuscles are formed of a delicate membrane, *stroma* (oikoid, Brücke), which contains the colouring matter (zoöid) hæmoglobulin, cholesterin, lecithin, and inorganic salts. The *stroma* is the colourless portion of the living blood corpuscle; it is insoluble in water and serum, and in sodium chloride solutions, but freely soluble in ether, chloroform, caustic soda, ammonia, and in solutions of the bile acids and urea. The stroma appears to combine with the hæmoglobin, and, so to speak, fixes it, but the union is very feeble, and very slight disturbing influences set free the colouring matter. The hæmoglobin in the living blood is combined with an alkali, probably potash, to keep it in solution, as otherwise it is very insoluble, and would crystallise out.

One hundred parts of human dried corpuscles contain hæmoglobin, 86·79; proteids, 12·24; lecithin, 0·72; cholesterin, 0·25 (HoppeSeyler). The corpuscles may be obtained for analysis by rapidly defibrinating blood, and then adding to it ten times its volume of a 10 per cent. solution of common salt, and setting aside in a cool place, when the corpuscles will be deposited. They are then to be collected on a filter, and washed thoroughly with sodium chloride solution till thoroughly free from serum. The mass is then placed in a weighed capsule and dried; the weight after drying represents the amount of corpuscles present in a certain quantity of blood.

(*a*) *Clinical determination of hæmoglobin.* — In disease the amount of hæmoglobin varies considerably. According to Quincke the following are some of the chief deviations observed : Taking 11·5 to 12·0 to represent the percentage amount of hæmoglobin in normal blood, then in a case of cirrhosis of liver, with epistaxis, it was 10·1 per cent. ; in chlorosis, 5·3, after taking iron, 9·92 ; in leucocythæmia, 5·8 ; in

Bright's disease, granular kidney, 8·5 and 10·60 ; with large fatty kidney, 10·30 and 10·70; in diabetes mellitus, 14·4 ; in phosphorus poisoning, 14·9. For clinical purposes the amount of hæmoglobin present in blood may be estimated by means of the number of blood corpuscles, and their depth of colour. This is done by counting first (by means of the hæmocyto-meter), the number of corpuscles in a small quantity of blood, and then comparing the colour of the same blood with a tint of a solution of hæmoglobin of known strength. For this purpose the instruments devised by Dr. Gowers * are the ones generally employed. First of all the number of the corpuscles has to be determined. This is done by the *hæmocytometer*. The proceeding is as follows : 5 cubic millimetres of blood are drawn into a capillary tube from a puncture, by means of a spear-pointed needle in the finger, and blown into a cup-like cell containing 995 cubic millimetres of solution of sodium sulphate (sodium sulphate and distilled water added till the specific gravity is 1·025), measured off by means of a specially graduated pipette. The contents of the cell are to be well mixed with a little spatula, and then one drop is to be placed in the cell-like depression, with ruled squares of $\frac{1}{10}$ millimetre each on the floor of the slide, to which springs are attached to secure a cover glass. The slide is then placed in the stage of a microscope, the lens focussed to bring the squares into view. A few minutes are allowed to let the corpuscles sink on to the squares, and then the number contained in 10 squares is counted ; this, multiplied by 10,000, gives the number of corpuscles in a cubic millimetre of blood. In the healthy blood of man the standard is 5,000,000 in each cubic millimetre. The depth of colour has now to be determined. To ascertain this two tubes have

* *Lancet*, Dec. 1st, 1877 ; and vol. ii., page 822, 1878.

to be employed; the one containing a solution of a
standard tint, equivalent to the dilution of 20 cubic
millemetres of blood in 2 cubic centimetres of water,
the other graduated so that 2 cubic centimetres give
100 degrees. The tubes are placed side by side, and
a sheet of white paper placed behind them. 20 cubic
millimetres of blood are then placed in the graduated
tube by means of a capillary pipette, a few drops of
distilled water being first placed in the bottom of the
tube. The mixture is to be kept constantly agitated
by a stirring rod. Distilled water is then gradually
added until the tint of the diluted blood is exactly
the same as that of the standard solution. In healthy
blood the tint is reached at 100 degrees of dilution.
In blood, poor in hæmoglobin, the degree of dilution
required to give the normal tint is less; so that 60,
70, or 80 degrees of dilution only may be required.
The result of these two observations is recorded as
the value of each corpuscle. Thus the blood from a
patient yields 4,000,000 corpuscles for the cubic
millimetre, instead of the normal standard, 5,000,000.
In other words, the percentage of corpuscles is 80
instead of 100, and the amount of dilution required to
bring the blood to the standard tint is 60 centimetres,
and if we take the percentage of corpuscles as a
denominator, and the percentage of colour as a nume-
rator, then $\frac{60}{80} = \frac{3}{4}$ as the value of a nominal corpuscle. The hæmacytometer is also used for deter-
mining the proportion of white to red corpuscles.

(b) *Spectroscopic characters of hæmoglobin and its
compounds.*—Whenever a tube* containing undiluted
blood is interposed between a source of light and a
spectroscope, the whole of the rays, with the exception
of the red, are obscured. On gradually diluting the

* Or more conveniently a hamatinometer, a vessel with two
parallel glass faces, one centimetre apart, so that the depth of the
stratum examined may be accurately determined.

blood the spectrum clears up, and the red, orange, and yellow become visible, and a portion of the green beyond E. But between D and E there is an intensely dark space. On further dilution this gradually clears up, leaving however two bands (Fig. 1, No. 1), one at a, near to D, with well-defined edges, and whose centre corresponds to 580—578 of the wave-line. The other, b, broader, less shaded and defined at the

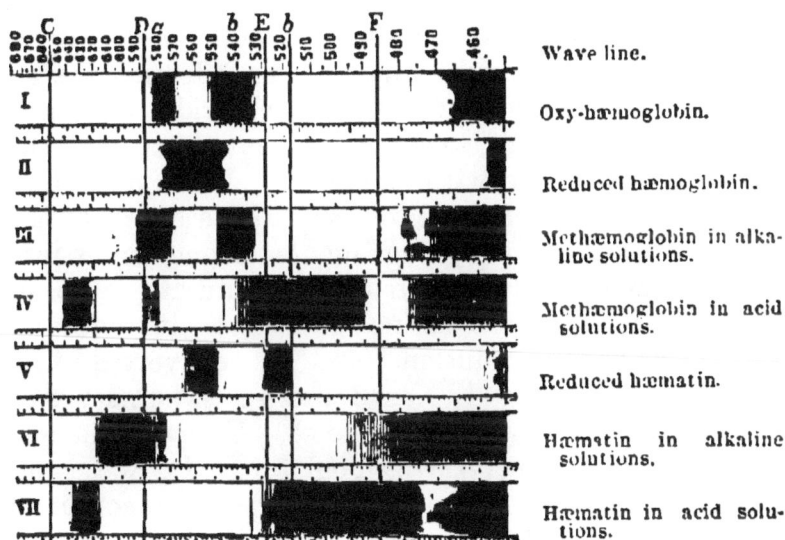

Fig. 1.—Spectrum of Hæmoglobin and its Compounds.

edges, near to E, whose centre corresponds to 540—543 of the wave-line. The two bands are characteristic of oxy-hæmoglobin. The space occupied by these bands is a measure of the amount of hæmoglobin present. Thus, in a solution one centimetre thick, and containing 0·8 per cent. hæmoglobin, the two bands are continuous from 595 to 520, whilst the green just beyond 518 is slightly visible; by increasing the strength of the solution the green disappears. With 0·35 per cent. the bands are separated; a extends from 588 to 568, and b 554 to 524.

With 0·09 per cent. solution *a* extends from 583—
571, and *b* from 550—532, with less than 0·01 per
cent. *a* is faint and extends between 583 and 575, *b* is
visible with difficulty, but Professor Gamgee gives its
position as 538—550. If we add certain reducing
agents to the blood, carefully excluding air, then the
oxy-hæmoglobin is deprived of its oxygen, and we
have the spectrum of *reduced hæmoglobin;* this gives
a single broad band extending from 595—540, diffuse
at its edges, and darkest between 560—550 (Fig. 1,
No. 2). The agent generally used for effecting the reduction is a solution of stannous chloride, to which
a small quantity of tartaric acid is added, and the
mixture neutralised with ammonia. When a solution
of hæmoglobin is exposed for long to the air, or to the
action of certain oxydising agents (ozone, potassium
permanganate, sodium nitrite, amyl nitrite, etc.) in
neutral or faintly alkaline solutions, it forms a
yellowish - brown solution, and is converted into
methæmoglobin (Fig. 1, Nos. 3 and 4). The nature of
this body is not well understood. Formerly it was
supposed to be per-oxy-hæmoglobin. It does not,
however, form oxy-hæmoglobin under the action of
reducing agents, but passes directly into the reduced
hæmoglobin. This and other considerations have
made recent writers view methæmoglobin as containing less oxygen and more iron than oxy-hæmoglobin,
but that the oxygen is in a more stable condition
than in the latter body. In nearly all solutions
containing methæmoglobin, the two bands of oxy-
hæmoglobin are generally visible. This is accounted
for by the presence of a portion of oxy-hæmoglobin
not being transformed into methæmoglobin. The
fluids of certain cysts, the deposits after the extravasation of blood into the cellular tissue, the blood after
inhalations of nitrous oxide, or the administration
of the nitrites, and the urine in hæmatinuria, give the

characteristic spectroscopic appearance of methæmo-globin. When a solution of hæmoglobin is treated with acetic acid, the two bands between D and E vanish, and instead we have a broad band between C and D, whose centre corresponds to wave length 625, this the spectrum of *acid hæmatin* (Fig. 1, No. 6). If the solution be now rendered strongly alkaline, this band shifts nearer D, with a centre corresponding to 610 of the wave line, whilst the blue end of the spectrum becomes more obscure. This is the spectrum of *alkaline hæmatin* (Fig. 1, No. 6). If to either acid an alkaline solution, the reducing agent (stannous chloride, tartaric acid, and ammonia) be added, no bands are seen in C, D, but instead two bands, not unlike those of oxy-hæmoglobin, are seen between D and E, one a broad band reaching from D half way to E, and the other a narrower band near E; this is the spectrum of *reduced hæmatin* (Fig. 1, No. 7). The spectrum of acid hæmatin is not unlike that of methæmoglobin in an acid solution, and has been mistaken for it; by rendering the solution alkaline the resemblance disappears.

(c) *The chemical properties of the colouring matter of the blood.*—The defibrinated blood of the rat, squirrel, or guinea-pig, received into a beaker surrounded with ice, and allowed to stand a short time, yields abundant and well-defined crystals of hæmoglobin, but with the blood of the ox, sheep, horse, or man, the separation in the crystalline form is not so readily obtained, and the following process must be followed :—100 centimetres of freshly drawn blood must be rapidly defibrinated and placed in a shallow vessel and treated with ten times its volume of 10 per cent. solution of sodium chloride, and set aside in a cold place (in summer placed in a refrigerator) ; when the corpuscles are deposited the supernatant liquid is decanted off, and the mass placed in a filter and washed repeatedly with sodium chloride solution till the washings are completely free from albumin. It is then agitated with a mixture of one volume of water to four volumes of ether, and allowed to stand; the etherial solution containing the fatty matter is then removed, and the red aqueous solution filtered into a beaker surrounded with ice, and alcohol carefully added till a precipitate begins to appear. After some hours crystals of hæmoglobin will form. The success of the proceeding depends on thoroughly removing the albumin and fatty matters, and conducting the whole process at a low temperature. The crystals in man form as prismatic needles, with dihedral summits, or in rhombic plates, Fig. 2, c and D. They are soluble in water and alkaline solution, and are insoluble in alcohol, chloroform, and ether. Acids decompose them with the formation of hæmatin.

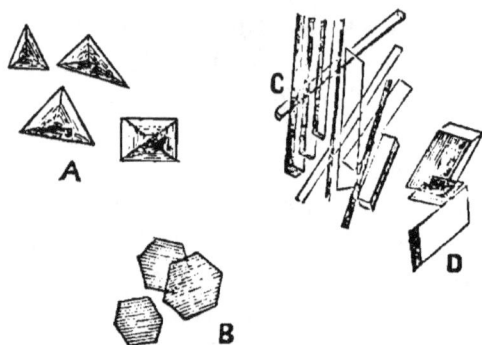

Fig. 2—Hæmoglobin Crystals.
A, Of guinea-pig; B, of squirrel; C, D, human.

Hoppe Seyler gives the composition of dried crystallised hæmoglobin as,

C	.	.	.	54·01
H	.	.	.	7·20
N	.	.	.	16·17
Fe	.	.	.	·42
S	.	.	.	·72
O	.	.	.	21·48

100·00

from which the formula $C_{6,0}H_{969}N_{151}FeS_3O_{179}$ may be
adduced ; and

$$
\begin{array}{rrcll}
12 & \times & 600 & C & = & 7200 \\
1 & \times & 960 & H & = & 960 \\
14 & \times & 154 & N & = & 2156 \\
56 & \times & 1 & Fe & = & 56 \\
32 & \times & 3 & S & = & 96 \\
16 & \times & 179 & O & = & 2864 \\
\hline
 & & & & & 13332
\end{array}
$$

gives the molecular weight 13332. And as one
molecule of hæmoglobin requires three molecules of
soda to form non-coagulable combinations, therefore
$\frac{13332}{3} = 4444$, or the equivalent weight of hæmoglobin.

The amount of oxygen linked with hæmoglobin to
form oxy-hæmoglobin, is 1·27 cubic centimetres of
oxygen at 0° C. and 1 metre pressure to 1 gramme of
hæmoglobin. Carbonic oxide and nitrous oxide dis-
place the oxygen in oxy-hæmoglobin, the solution ac-
quiring a bluish tint, but the spectrum is little altered,
and is not affected by reducing agents, whilst the
combination seems more stable than that of oxygen
with hæmoglobin. *Hæmatin* $C_{68}H_{70}N_8Fe_2O_{10}$ is best
prepared by mixing defibrinated blood with a strong
solution of potassium carbonate, till the liquid adhering to the separated coagulum becomes colourless.
The coagulum is then dried at 50° C. and digested for
some days in absolute alcohol ; the alcoholic solution
after concentration will deposit rhombic crystals.
Hæmatin crystals are of a bluish-black colour with a
metallic lustre, becoming brown on trituration. They
are insoluble in water, alcohol, ether, and chloroform,
but soluble in acids and alkalies ; the acid alcoholic
solutions are monochromatic, having a brown colour ;
the alkaline solutions are dichromatic, and have an

G

olive-green colour and dark red in the thicker layers.
The crystals support a temperature of 180° C., but
above that they carbonise.   When treated with
strong sulphuric acid hæmatin is deprived of its iron ;
this substance, to which Hoppe Seyler has assigned
the name of *hæmatopopphyrin* (thus, $C_{68}H_{70}N_8O_{10}Fe_2 +$
$4SO_4H_2 + O_3 = C_{68}H_{70}N_8O_{10}(SO_4H_2) + 2FeSO_4 + 2H_2O)$,
gives to the spectrum a dark band midway between
D and E, and a narrow band between C and D
(nearer D); the spectrum also is deeply shaded
between D and E.   If a solution of hæmoglobin
reduced by hydrogen is decomposed by sulphuric
acid, a substance is formed which Hoppe Seyler has
named *hæmocromogen*, and which by oxydation yields
hæmatin.   This body when in alkaline solution is
identical with reduced hæmatin, whilst in acid solution
its spectrum gives a combination of those of acid hæmatin and hæmatopopphyrin.

Hæmin $C_{68}H_{72}N_2Fe_2O_{10}Cl_2$, or hæmatin hydrochlo-
ride ; if a small quantity of blood is rubbed up with
sodium chloride and boiled for a few minutes with
glacial acetic acid, and the mixture evaporated to dryness, in the residue, mixed
with colourless crystals of sodium chloride
and sodium acetate, will be found rhombic
tablets of hæmin (Fig. 3); which are of a
bluish - red colour when viewed by reflected, and brownish - red by transmitted, light.
They are insoluble in hot and cold water, in
alcohol and ether.   Soluble in alkaline solutions.
All acids, with the exception of hydrochloric and
acetic acid, decompose them.   Heated to 200° C.
they undergo decomposition, evolving fumes of
prussic acid, and leaving a residue of oxide of iron.

Fig. 3.—
Hæmin
Crystals.

Hæmatoïdin $C_{15}H_{18}N_2O_3$ (Robin and Verdeil).
Under this name Virchow described certain red
crystals found in clots of old extravasations, as in

apoplectic clots, corpora lutea, etc. They have been considered to be identical with bilirubin, but the question is not yet decided. They can be obtained from the corpora lutea by rubbing them up with pounded glass, agitating frequently with chloroform. After standing some time the chloroform solution is poured off and allowed to evaporate. The crystals thus obtained are red by transmitted, and green by reflected, light ; they occur in rhombic plates, and are soluble in chloroform and ether, but insoluble in alcohol, water, and alkali.

96. **Examination of blood stains.**—The surface or substance of the material must be scraped or cut into small fragments and digested in as little distilled water as possible. Of the reddish fluid examine (1) under microscope for blood corpuscles ; (2) placed in deep narrow cell and examined by a spectroscopic eye-piece with a low power of microscope, for bands of hæmoglobin ; (3) add a few drops of glacial acetic acid and a small quantity of sodium chloride evaporated to dryness at 40° C. and 50° C., and examine residue for hæmin crystals ; (4) place a drachm of tincture of guiacum in a test tube and add a drop of the solution, then float on surface etherial solution of hydrogen peroxide ; if blood, a blue ring will form at junction of the etherial solution and the guiacum. (N.B.—Other substances besides blood give this reaction with guiacum.)

97. The **colourless corpuscles.**—These are the white and the intermediate. The former are nucleated masses of protoplasm of granular appearance, possessing the power of amœboid movement. They are larger than the red corpuscles, the diameter being $\frac{1}{2500}$th of an inch. The intermediate or hæmatoblasts are smaller than the white corpuscles, and the nucleus is more obscured by granules. Treated with acetic acid the colourless corpuscles swell up,

rendering the nuclei more distinct. Touched with a solution of iodine and potassium iodide, the body of the corpuscles is stained a mahogany-brown, indicating the presence of glycogen. The proportion of white corpuscles to red may be stated as 1 to 340—350, but the proportion varies considerably under different physiological conditions, being increased by fasting and diminished by food. In leucocythæmia the proportion of white corpuscles to red is greatly increased, so as often to amount to one-eighth or one-tenth of the coloured corpuscles; they are of the same shape as in normal blood, but often smaller. On standing, delicate crystals often form in them; these are supposed to be a phosphate of a proteid base. In progressive pernicious anæmia the white corpuscles are not absolutely increased. In addition to the white corpuscles in leucæmia and pernicious anæmia, there are often nucleated coloured corpuscles similar to those found in the blood of the human embryo; in these cases the marrow of the bones is generally affected, whether as a primary or secondary condition is still undetermined.

98. **Blood serum,** as it separates from coagulated blood, is clear fluid of a straw colour, sp. gravity 1·025—1·028; its alkaline reaction is higher than an equal bulk of blood. It contains, in addition to fatty matters, extractives, and salt, the other proteids of the blood, viz., paraglobulin and serum albumin. *Paraglobulin,* serum globulin, or fibrinoplastic substance, is precipitated from blood serum by adding to it magnesium sulphate to complete saturation when it is precipitated. Hammarsten by this method has obtained 3·103 parts of paraglobulin in 100 parts of blood. It is said to pass easily through animal membranes, which serum albumin does not. It is, therefore, not improbable that some cases of temporary albuminuria may be due to the passage of this body

into the urine, and not altogether to the presence of
serum albumin, as generally supposed.   Paraglobulin
with fibrinogen in the presence of the fibrin ferment
forms fibrin ; fibrinogen, however, is only found in the
blood plasma.   The *serum albumin* can be obtained
after the removal of the paraglobulin, by gently con-
centrating the filtrate at a low temperature, 30° C.
to 35° C., and then placing it on a dialyser to remove
saline impurities ; the solution will then give the reactions described (§ 20).

99. **Fatty matters.** — A definite quantity of
blood is evaporated to dryness over a water-bath, and
when completely dry the mass is broken up and
carefully triturated in a mortar, and the powder
thoroughly exhausted with boiling ether.   The etherial
solution is then evaporated in a weighed platinum
capsule.   The increased weight of the capsule gives
the amount of fats in the quantity of blood examined.
Normal blood yields from ·18 to ·2 per cent. of fatty
matter, which consists of saponifiable fats, lecithin,
and cholesterin, the two latter being chiefly, if not
altogether, derived from the corpuscles.   To obtain these
bodies separately, the residue in the platinum capsule
must be again treated with ether and boiled with
baryta water ; the fatty acids are thus converted into
baryta soap, and can be removed by filtration.   The
precipitate is then treated with absolute ether, which
removes the cholesterin from the soaps, and is evaporated on a weighed filter and the amount of cholesterin
ascertained.   The filtrate is then divided into two portions.   One is evaporated till it is dry, and is then
extracted with absolute alcohol ; the neurin, one of the
products of the decomposition of lecithin, is then precipitated by platino-chloride solution, as neurin platino-
chloride.   The other portion of the filtrate is evaporated
to dryness, and fused with sodium hydrate and nitre,
and the residue dissolved in water and nitric acid added

in excess ; a solution of ammonium molybdate gives a yellow deposit, which is to be dissolved in ammonia. To this, solution of magnesium sulphate and ammonium chloride is added, and on spontaneous evaporation crystals of ammonio-magnesium phosphate will form. The crystals are collected in a platinum capsule, and ignited as directed for the determination of magnesia (§ 85). Then since 100 parts of the pyrophosphate are equivalent to 764·8 parts of lecithin, a calculation of the amount of lecithin present can be made. Thus if the weight of the pyrophosphate amounts to ·026 grammes, then the lecithin will amount to 0·19764 grammes, but as only half the filtrate was taken, then the total lecithin in the fatty residue will be 0·39528. Now supposing the amount of blood examined for fatty matter yielded a residue of total fats of 1·9 per cent., then by deducting the weight of the cholesterin, which on weighing has been found to be ·08 grammes, and deducting the weight of the lecithin as found above, we have the weight of the saponifiable fat. So that 100 grammes of blood have yielded 1·9 parts of mixed fats, of which ·08 is cholesterin, ·395 is lecithin, and the remainder 1·425 saponifiable fat. The amount of fat in the blood is increased after a full meal and diminished during fasting. In diabetes the blood sometimes assumes a lactescent appearance (*lipæmia*), due to the presence of excess of fatty matter. This condition, which is rare, is generally met with in cases that have run an acute course, and terminate in a peculiar form of coma. (*See* Acetonæmia.) Cholesterin, it is said, is also to be found in excess in the blood in cases of liver disease. (*See* Cholesteræmia.)

100. **Extractives** consist chiefly of urea, glucose, kreatin, hypoxanthine, and uric acid. (1) *Urea* can be determined either (*a*) by diluting 50 grms. of blood with 200 of distilled water and adding five drops of sulphuric acid, and boiling the mixture. Filter.

To the filtrate add a saturated solution of baryta hydrate (2 volumes) and barium nitrate (1 volume) to throw down sulphates and phosphates, and add a few drops of strong solution of silver nitrate to precipitate chlorides. Filter. Add cautiously to filtrate some sulphuric acid till there is an acid reaction. Filter. Evaporate filtrate to consistence of syrup, and add 50 cc. of absolute alcohol. Filter. Evaporate alcoholic solution and dissolve residue in 50 cc. of water. To this add cautiously from a Mohr's burette, drop by drop, Liebig's solution of mercuric nitrate (*see* Urine,§ 110), diluted with an equal quantity of distilled water, stirring after each addition, and removing a drop of the mixture to a plaster of Paris slab moistened with drops of sodic carbonate solution. When a yellow stain is given to one of these drops, no more mercuric solution is to be added. As 1 cc. of the mercuric solution is equivalent to ·005 gramme of urea, the number of centimetres of the solution used indicates the amount of urea in 50 cc. of blood. (*b*) Defibrinate* 20 cc. of blood, place it on a parchment paper dialyser, and spread it over it so as to form a thin layer. Float it in a vessel containing 50 cc. of absolute alcohol. From time to time add a very little distilled water to keep the mass on dialyser moist; continue the process for twelve hours. Treat the diffusate with an equal bulk of solution of concentrated oxalic acid, and evaporate to dryness. To residue add some naphtha petroleum to remove fatty matters. Dissolve the purified residue in a little water, and add barium carbonate. Evaporate. Treat the residue with boiling alcohol and filter. Concentrate filtrate, from which on cooling urea will crystallise out. The percentage amount of urea in healthy blood ranges from ·025 to ·035 grms. In Bright's disease, acute yellow atrophy

* Haycraft's method, from Professor Gamgee's work.

of the liver, and in gout, the amount is considerably increased.

(2) *Glucose.* Dr. Pavy's process gives the most reliable results hitherto obtained. It is as follows :— Forty grammes of sulphate of soda in small crystals are weighed out in a beaker of about 200 cc. capacity. About 20 cc. of the blood intended for analysis is then poured upon the crystals, and the beaker and its contents again carefully weighed. The blood and crystals are well stirred together with a glass rod, and about 30 cc. of a hot concentrated solution of sulphate of soda added. The beaker is placed over a flame guarded with wire gauze, and the contents heated until a thoroughly formed coagulum is seen to be suspended in a clear colourless liquid ; to attain which actual boiling for a short time is required. The liquid has now to be separated from the coagulum, and the latter washed to remove all the sugar. This is done by first pouring off the liquid through a piece of muslin resting in a funnel into another beaker of rather larger capacity. Some of the hot concentrated solution of sulphate of soda is then poured on the coagulum, well stirred up with it, and the whole thrown on the piece of muslin. By squeezing, the liquid is expressed; and to secure that no sugar is left behind, the coagulum is returned to the beaker, and the process of washing and squeezing repeated. The liquid thus obtained may be fairly regarded as containing all the sugar that existed in the blood. From the coarse kind of filtration and squeezing employed, it is slightly turbid, and requires to be thoroughly boiled to prepare it for filtration through ordinary filter-paper. A perfectly clear liquid runs through, and to complete this part of the operation the beaker that has been used, and the filter-paper, are washed with some of the concentrated solution of sulphate of soda before referred to.

The next step is boiling with the copper-test solution. The liquid is again placed over a flame, and brought to a state of ebullition. A sufficient quantity of the copper solution to leave some in excess is now poured in, and, from the time of recommencement of boiling, brisk ebullition is allowed to continue for a period of one minute. As regards the amount of copper solution to be used, although 10 cc. of the test, as ordinarily made, are found to suffice for 20 cc. of the blood of animals in a natural state, yet it is well to employ from 20 to 30 cc. to secure that it is thoroughly in excess. Where, from any circumstance, larger quantities of sugar exist in the blood, more in proportion of the test must, of course, be used.

The precipitated suboxide of copper has now to be separated from the excess of copper solution. Experience shows that filtration through filter-paper cannot be resorted to for the purpose. A material, however, which has somewhat recently been introduced, viz., glass-wool, fully furnishes what is wanted. Properly packed in the neck of a funnel, it permits filtration to be effectively and easily performed. The filtrate should always be carefully examined, to see if the plug has been sufficiently tightly packed to keep the whole of the precipitate back. Should the crystallisation of the sulphate of soda in this or the preceding filtration interfere with the continuance of the operation, the funnel may be placed over a beaker holding some liquor kept in a state of ebullition, or heat may be applied in any other way.

The suboxide having been collected, and washing with distilled water performed, it is returned to the beaker in which the reduction was effected, to secure that none of the precipitate that may have been adhering to the sides of the vessel is lost. The plug is pushed with a glass rod from the neck of the funnel held in an inverted position over the beaker,

and the funnel washed and its surface cleaned from
all adhering precipitate. Now the suboxide is in a
fit state to dissolve, and, after the addition of a few
drops of peroxide of hydrogen, a very small quantity
of nitric acid (a few drops only) is sufficient to lead
to instantaneous solution, and, after boiling, to decompose the excess of peroxide of hydrogen, the contents
of the beaker, consisting of filter-plug and dissolved
precipitate, are poured into a funnel containing a
loose plug of glass-wool, to obtain the liquid in a
separate form. The requisite washing with distilled
water having been performed, there only remains the
final stage of the process to be accomplished.

The liquid to be now dealt with contains the
copper in the form of nitrate, which experiment has
shown to be the most suitable for yielding a pure
metallic deposit by galvanic action. For the purpose
of collecting the deposit, a cylinder of platinum foil,
soldered to a platinum wire for hooking on to the
negative pole of the battery, is employed. This is
immersed in the liquid so as nearly to touch the
bottom of the vessel, and inserted within it is a spiral
coil of platinum wire, made to constitute the positive
pole of the battery. In order to secure a good continuous connection, the platinum spiral is closely
bound to the copper conducting-wire of the battery,
and the other pole is provided with a platinum hook
for the suspension of the cylinder. At the end of
twenty-four hours' exposure to galvanic action the
weight of the cylinder with the deposited copper is
taken. The cylinder is lifted quickly out of the
liquid, and instantly plunged, first into distilled
water, and then into spirit, the latter being used to
avoid the occurrence of oxydation of the copper in
drying. After drying by suspension in a water-
oven, the process of weighing is performed, and it is
hardly necessary to say that a delicate chemical

balance is required for the purpose. The weight of
the cylinder being known and subtracted gives the
weight of the copper that has been thrown down. In
the case of an analysis of blood containing an
ordinary amount of sugar, and therefore yielding a
limited amount of copper to be deposited, twenty-
four hours have usually been found to suffice for the
completion of the operation ; but it is necessary there
should be no uncertainty upon this point, and, to
secure this, the following course of procedure should
be adopted. After the weighing has been effected,
the deposited copper is dissolved off by immersion of
the cylinder in nitric acid, and the cylinder then
returned into the liquid to see if any fresh deposit
occurs. If, after some hours, no copper tint is seen,
the operation may be regarded as completed ; but, if
more deposit has occurred, the immersion must be
continued, and another weighing performed, and this
repeated till the platinum surface remains untinted.

The galvanic action requires to be steadily and
continuously maintained, and a modification of Fuller's
mercury-bichromate battery has been found to answer
best for use. The arrangement that was employed
in Dr. Pavy's experiments consisted of an outer cell
provided with two carbon plates, and charged with bichromate of potash dissolved to saturation in dilute sulphuric acid. Into the inner porous cell a little mercury
is poured, and it is then filled up with water. An
amalgamated zinc rod is inserted, and dips down into
the layer of mercury at the bottom. This battery, it
is found, gives a steady current, and, used every day,
will remain in good working order for at least a
fortnight, all that is necessary being to pour out the
liquid in the porous cell when it has become green
from reduction of the diffused bichromate solution,
and replace it with water. Attention is, of course,
necessary to secure that the proper battery-power

exists to effect the deposition of the copper, and, when the current becomes weak, the zinc rod must be cleaned, and the bichromate of potash solution replenished.

When sugar is boiled with the copper solution, the change occurring stands in the relation of one atom of the former to five atoms of cupric oxide. One atom of sugar is oxydised by, or reduces, five atoms of cupric oxide. This is the foundation of the action involved in the operation of the test, and the calculation of the amount of sugar present is made accordingly. Taking 63·4 as the atomic weight of copper, and 180 as that of glucose $C_6H_{12}O_6$, 317 parts of copper will stand equivalent to 180 parts of glucose. Thus, one part of copper corresponds to ·5678 of glucose, and, in calculating the amount of sugar in the blood analysed, the weight of the copper deposited has only to be multiplied by ·5678 to give its equivalent in glucose. The quantity of sugar in the amount of blood taken for analysis being thus determined, the required information is supplied for expressing the proportion for 1000 parts.

(3) *Kreatin* is found in only very small quantities in blood. In order, therefore, to obtain it in any appreciable amount, large quantities of blood are required for analysis. Take 1000 grms. of blood, and remove albuminous constituents, phosphates, sulphates, and chlorides, as described for the determination of urea. The clear filtrate is then precipitated by a solution of basic lead acetate. Filter. The filtrate is then decomposed by sulphuretted hydrogen to remove the lead, and filtered through animal charcoal. The filtrate is then evaporated gently to a syrupy consistence, and treated with twice its volume with alcohol, from which mixture crystals of kreatin (§ 62) will deposit. [The student who wishes to obtain kreatin as a specimen can readily obtain it by dissolving 5 grms. of meat extract in 50 cubic

centimetres of water, adding basic lead acetate, and
proceeding with the filtrate as directed above.]

(4) *Hypoxanthin* is found in the juice of flesh, in
the spleen, and in the thymus and thyroid glands.   In
leucocythaemia it is found in appreciable quantities in
the blood.   To determine it, take 100 grms. of blood,
and separate the albuminous constituent, the phosphate
sulphate, and chlorides.   The aqueous solution is
then treated with ammoniacal solution of silver
nitrate, as directed.   (*See* Urine, § 125.)

(5) *Uric acid* cannot be obtained from healthy
blood in quantities sufficient for identification.   In
gout, however, this substance can be obtained for
examination by placing about two drachms of serum
(obtained from a blister) in a large watch-glass, and
adding to it, by means of a glass rod, acetic acid
till it has a decided acid reaction.   Place in the mixture a fibre of coarse linen, and allow it to stand till
the contents of the watch-glass become gelatinous ;
then take out the fibre, and crystals of uric acid will
be found adhering to it (§ 111).

101. **Salts.**— The inorganic residue of blood is
generally determined by incinerating a given quantity
of blood in a muffle furnace, or over a Bunsen lamp, till
the ash is nearly colourless, and making a quantitative
estimation of the acids and bases present.   This
method is sufficiently accurate when we wish to determine only the bases ; but, in the case of the phosphates
and sulphates, it is liable to error, since the phosphorus contained in the phosphorised fats, and the
sulphur from the proteids, become oxydised, and combines with the bases.   It is therefore best to calculate
the bases directly from the ash, according to directions
§§ 85 and 88, and make a separate determination of
the acids directly from the blood.   This can be done
by placing a small quantity of blood, dried at a low
temperature and completely pulverised, and then

moistened with water, on a dialyser, and floating it in a small quantity of distilled water, till what comes from the dialyser when placed in a fresh quantity of distilled water no longer gives a haze with solution of silver nitrate. The diffusate contains both the oxydised and unoxydised inorganic residue, and the former can be calculated volumetrically by the processes given for the estimation of phosphoric acid (§ 113), hydrochloric acid (§ 114), and sulphuric acid (§ 115). (*See* Urine.)   An analysis so conducted yielded the following results : 100 parts of blood the ash yielded, lime, ·011 ; magnesia, ·006 ; potash, ·034 ; soda, ·374 ; iron oxide, ·045 ; phosphoric acid, ·112 ; sulphuric acid, ·035 ; hydrochloric acid, ·285. This analysis shows the preponderance of soda over potash, a preponderance which is still more marked if only the serum be taken for analysis.   On the other hand, the potassium salts are more abundant in the corpuscles (§ 10).   The chlorides, again, are more abundant in the serum, the phosphates in the corpuscles. With regard to the combination existing between the bases and the acids, the soda combines with chlorine to form 0·53 per cent. of the ash, and with phosphoric acid as neutral phosphate Na_2HPO_4, whilst ·15 parts of soda are combined with carbonic acid as normal carbonate Na_2HCO_3, and as acid carbonate NaH_2CO_3.   It is owing to the decomposition resulting between the neutral phosphate of soda and the acid carbonate that acid secretions are separated from the alkaline blood (§ 9), a fact I showed experimentally in 1874,* and which has subsequently been confirmed by other observers.   The phosphoric acid is chiefly combined with potash in the corpuscles, and with lime and soda in the serum.

102. **Toxic conditions of the blood.**—When any impediment is offered to the excretion of the

* *Lancet,* July, 1874.

effete materials of the organism by the natural channels, these accumulate in the blood, and induce symptoms of great gravity. When, for instance, the process of aëration of the blood by means of the lungs is checked to any great extent, carbonic acid accumulates in the blood, and the condition of *asphyxia* is induced. This, when the blockage of the air-passages is complete, proves rapidly fatal. In less severe forms of obstruction, the condition is marked by dyspnœa, and a more or less livid and purplish condition (*cyanosis*) of the skin and visible mucous membranes, due to the presence of imperfectly aërated blood in the capillaries. Cyanosis is the most conspicuous symptom of congenital deformity of the heart. Insufficiency of the tricuspid valve, causing as it does great engorgement of the veins of the systemic circulation, is generally accompanied by marked cyanosis. In emphysema, especially in the later stages, when the right side of the heart ceases to compensate by its hypertrophy for the circulatory impediment, the cyanosis becomes extremely intense; even in the early stages, with the loss of the interalveolar septa and disappearance of the capillaries, the imperfect aëration of the blood is manifest in the leaden hue of the patient's countenance. In disease affecting the left side of the heart, cyanosis is not a marked symptom. Niemeyer ingeniously accounts for this by the supposition that in valvular affections of the left heart the pulmonary circulation is surcharged with blood, whilst the quantity of blood in the systemic circulation is abnormally small ; whilst in emphysema, where many pulmonary capillaries have perished, and in diseases of the tricuspid and pulmonary valves, it is the systemic circulation which is overloaded, whilst the pulmonary contains too little blood for effectual aëration. In diseases accompanied by diminution in the number and colour value of the red corpuscles, in addition to the pallor, there is

usually a well-marked leaden hue; this is particularly
noticeable in patients suffering from scurvy.

In diseases of the kidney, when the excretion of the
solid matters of the urine is considerably diminished,
a condition known as *uræmia* often supervenes. This
is generally attended by a drowsy condition, more or
less intense, with the frequent recurrence of convulsions of an epileptic character. The urine is greatly
diminished; but if it should be increased, as it is in
some rare cases, it is of very low specific gravity.
Numerous views have been advanced to account for
the symptom. Some consider it due to the poisonous
action *per se* of retained urea; others, that the urea
is decomposed into ammonium carbonate, and the
blood is poisoned with ammonia; while others hold
that excess of urea in the blood causes the water of
that fluid more readily to transude through the capillaries, and thus cause œdema of the brain. Against
these views it may be urged that urea may be injected
in large quantities into the veins of animals without
inducing uræmia; that no ammoniacal odour is perceptible in the health of persons suffering from this
condition, whilst in typhus fever, in which ammonia
has been recognised in the breath by many reliable
observers, the coma is not of uræmic character. And
lastly, uræmic convulsions occur in some cases without
there being evidence of any marked diminution in the
excretion of urea, as in puerperal convulsions. In
this condition a considerable quantity of albumin is
passed in to the urine, amounting, in three examinations I have made, from six to eleven grms. in the
twenty-four hours; the urea, on the other hand, was
only slightly below the normal rate of excretion. Considering uræmic convulsions occur chiefly in cases
where the drain of albumin has been considerable,
either in large quantities as in acute nephritis, or in
small quantities but of long continuance, as in chronic

renal disease, I think we may reasonably attribute the condition in some measure to the withdrawal of the nutritive matter of the blood, or, at least, to the altered percentage relationship between it and the effete (extractive) materials. One point is clear, that venesection often proves of immense service in this condition, - apparently by altering the percentage composition. In one case of uræmic convulsions, in a lady seven months advanced in pregnancy, whose urine I examined at the request of Dr. John Williams, I found in the urine passed during twelve hours immediately preceding venesection, 5·1 grammes of albumin, whilst the urine passed in the twelve hours immediately after venesection only contained 2·3 grammes. Free purgation, by relieving the tension in the renal capillaries, has much the same effect, though to a less extent. A patient in a state of deep coma, which had been preceded by convulsions, had a drop of croton oil administered, his urine at the time containing about one-fifth albumin; after free purgation he recovered consciousness for several hours and the urine was found on examination to contain only one-twelfth albumin.

In diseases of the liver, when any obstruction is offered to the onward passage of the bile, re-absorption takes place, and the bile passes into the blood and is deposited in the tissues, giving rise to the phenomenon of *jaundice*. (*See* Bile, chapter v.)

In certain forms of liver disease, particularly cirrhosis, in which there is a great destruction of liver cells, the termination is by coma, and has been attributed to the retention of cholesterin in the blood (*cholesteræmia*). We have no evidence, however, to show that cholesterin has any toxic influence. Krusenstern injected from ·005 to ·045 gramme of cholesterin daily into the veins of dogs, and found the animals unaffected. Pagès arrived at the same

H .

results.* Looking at the question from the purely clinical side, one would expect cholesteræmia in all cases of jaundice where there was considerable resorption of bile. If the liver is the seat, as is generally held, of the destruction of the blood corpuscles, which contain nearly all the cholesterin found in the blood in the normal state, we may naturally suppose the cholesterin of the bile re-absorbed in a free state into the blood, would rapidly induce cholesteræmia; but, as a matter of fact, the phenomena attendant of this condition are not common to jaundice, due simply to obstructions, however complete, unless there be also considerable destruction of liver tissue and disease of the kidneys. Again, in acute yellow atrophy, where we have rapid destruction of the liver cells, and where jaundice is comparatively slight, the coma is preceded and attended with symptoms much resembling those witnessed when acids, or phosphorus, are injected into the veins of animals, or in patients dying from acute diabetic coma (acetonæmia). From these considerations I venture to think that the coma attendant upon hepatic disease in which there is considerable destruction of liver cells, is not due to the accumulation of cholesterin, which, in many cases of ordinary jaundice, increases four or five per cent. (Frerichs) without inducing the condition, but to a general increase of the excretory matters in the blood, and this supposition is strengthened by the fact that the condition is never witnessed unless there is also some impairment as well of the renal functions.

In diabetic urine there is often found a body that gives with ferric chloride a deep red reaction, and corresponds, in many respects, with acetone, or acetone-yielding bodies. As this substance has been found most frequently† in the urine of patients dying rapidly

---

* *Journal de l'Anatomie et de la Physiologie*, 1875.
† Acetone can be obtained from most diabetic urines by distillation with hydrochloric acid.

of acute diabetic coma, it has been supposed that this phenomena is due to the presence of acetone in the blood (*acetonæmia*). Free acetone, however, has not, I believe, ever been obtained in a free state from freshly-drawn diabetic blood, but there is little doubt that a body readily yielding acetone can be separated from the blood of such patients. The nature of this body is a matter of some dispute; most writers consider it to be ethyl diacetate, which, by decomposition, yields equal molecules of acetone and alcohol. In this case, by distillation we ought to obtain equal amounts of acetone and alcohol; indeed, as acetone is the most volatile, it should be found in less amount. Some recent observations,* however, have shown, on the contrary, that when diabetic urine is distilled, the acetone is considerably in excess of the alcohol, and hence it has been surmised that the body is some compound of aceto-acetic acid. When acetone, or the acetone-yielding body, is found in the urine, there is as well a peculiar odour, like that of acetone, exhaled by the breath, and the urine likewise. This odour is generally particularly noticed as preceding acute diabetic coma, though when that condition is thoroughly established, it often disappears. After death, a lactescent or milky appearance is sometimes noted, with fatty changes of the liver and other viscera, though these conditions are far from being invariable. Acute diabetic coma or acetonæmia is distinguished from the more common but less characteristic form of coma, which usually terminates the disease, in the suddenness of its onset, the gastric disturbance, as shown by the acute epigastric pain, the vomiting (sometimes of blood), and more rarely, frequent purging; the peculiar noisy delirium, the panting respiration like an animal with both vagi cut, the fluctuations in the

* Deichmuller, Annalen, 209, 22—30. B. Tollens, Annalen, 209, 30—38.

rapidity of the pulse, which maintain till the coma is well established, when the pulse continues intensely rapid and small. These are symptoms which come on in rapid succession, and present a parallel to those which occur in acute yellow atrophy and phosphorus poisoning, or the injection of the bile acids or mineral acids into the blood of animals. The origin of the acetone-yielding body has not been determined, but it is probable that it is derived from some metamorphosis of the sugar in the blood, forming alcohol and the products of alcoholic fermentation. When we consider the highly acid character of the urine in diabetes, and the fact that symptoms of acute diabetic coma are not unlike those produced in animals poisoned by the injection of acid into their veins, it is not improbable that the body concerned is of an acid nature—aceto-acetic acid, as has been suggested. This body is probably present in the blood of most diabetics, in small quantities, and by its elimination from the urine gives that secretion the highly acid reaction met with in diabetes. When, however, it is formed in excessive quantities, or its elimination is interfered with, it accumulates in the blood, giving rise to the condition known as acetonæmia. The presence of acetone, or the acetone-yielding body, can be demonstrated in urine by the deep red coloration given by ferric chloride, which disappears on the addition of hydrochloric acid, and by the reaction with iodide of potassium. This last I have endeavoured to make available as a clinical test, as follows : About a drachm of liquor potassæ, containing twenty grains of iodide of potassium, is placed in a test tube and boiled ; a drachm of the suspected urine is then carefully floated on the surface. Where the urine comes in contact with the hot alkaline solution, a ring of phosphates is formed, and after a few minutes, if acetone or its allies are present, the ring will become yellow, and studded

with yellow points of iodoform ; these in time will sink
through the ring of phosphates, and become deposited
at the bottom of the test tube.

103. **Chyle and lymph.**— In examining these
fluids the same steps are to be taken as in the examination of blood, and the same methods employed for the
determination of the specific gravity, the reaction, the
amount of fibrin and other proteids, the fatty matter,
the extractives, and the salts. The chyle derived from
the intestinal lacteals does not contain fibrin, and
consequently does not separate into clot and serum.
If the animal has been fasting, the chyle loses its creamy
appearance and becomes more transparent and of a
yellowish colour. 100 parts of chyle yield, when
taken from the lacteals in full digestion : water, 91·8 ;
solids, 8·2 ; fibrin, 0·2 ; proteids, 3·5 ; fats, 3·3 ; extractives, 0·4 ; salts, 0·8 ; when fasting : water, 96·8 ; solids,
·38 ; fibrin, ·09 ; proteids, 2·30 ; fat, ·04 ; extractives,
·28 ; salts, ·49. The proteids consist of serum albumin,
paraglobulin, fibrinogen, and peptones. The fatty
matters, which consist of minute spherical globules,
and form what Gulliver called the molecular base of
the chyle, are a mixture of saponifiable fats, cholesterin, and lecithin (100 parts of ether extract yield
81· saponifiable fat ; 7·5 parts lecithin ; 11·3 parts
cholesterin ; Hoppe Seyler). The extractives are urea,
and glucose, whilst leucin and tyrosin are frequently
obtained. The constitution of the ash resembles that
of the blood, the sodium having a preponderance over
the potassium salts. Lymph is a clear, colourless, or
straw-coloured fluid, of slightly alkaline reaction. In
composition lymph closely resembles chyle, differing
chiefly in the smaller proportion of fibrin and fatty
matters it contains.

104. **Milk.**—The following gives the average
composition of human milk in 100 parts : water, 86·86 ;
solids, 13·2 ; proteids (chiefly casein) 2·93, butter 3·78,

sugar of milk 5·83, extractives 0·25, salts 0·35. It varies, however, considerably with the character of the food and other physiological conditions. Of the constituents, however, the casein and sugar are the most constant, whilst the fatty matters show a wide range of variability. Analyses have shown that the composition of milk varies with the age of the infant. Thus the casein is at its lowest at the commencement of suckling, and then gradually rises till it attains a fixed proportion, whilst the sugar is at its maximum at the commencement and subsequently diminishes. This is important in relation to the artificial feeding of infants. A comparison of human milk with cows' milk shows that the latter contains more casein but less sugar, whilst asses' milk, though poorer in casein than human milk, is quite as rich in sugar. The *colostrum* or the milk, passed during the first week after delivery, has a more alkaline reaction and higher specific gravity than ordinary milk, and is richer in casein and fatty matters. In certain morbid conditions the composition of the milk may be altered, though observations on this point are much wanting. In pyrexia the secretion is diminished, or may be quite suppressed, and the quantity of solids, chiefly the fats and sugar, fall considerably. After undue excitement, shock, mental emotion, the milk has been noticed to have an acid reaction, or to become so shortly after secretion. In some cases rapid decomposition sets in with the evolution of sulphuretted hydrogen. Certain substances taken as food or medicine find their way into the mammary secretion and for a time render it unfit for use. During the latter stage of pregnancy and the earlier period of lactation sugar sometimes appears in the urine; this must not be confounded with the grape sugar of true diabetes, but is the milk sugar, lactose (F. Hofmeister "Ueber Lactosurie, *Zeitsch. f. Phys. Chem.*," i. § 101), and can be distinguished by its

physical and chemical characters (§ 13). Milk sometimes assumes a distinctly blue appearance; this is due
to the development of bacteria. When milk is added
to an artificial solution of gastric juice it becomes
curdled, and this curd is gradually dissolved as digestion proceeds. This curdling is not due to the acid of
the gastric juice, but to some ferment which sets
up lactic acid fermentation in the milk sugar, as
is evident by the fact that if the artificial juice is
neutralised before the addition of the milk, the
curdling takes place just the same. The specific
gravity of the milk and its alkaline reaction are
determined by the same methods as described for
blood. The casein is separated by adding a few drops
of acetic acid, and boiling. Collecting the curd and
drying it, rubbing up the dried residue in a glass
mortar and frequently extracting with ether to remove
the fatty matter; placing the purified residue on a
weighed filter and drying it. The fatty matter,
extractives, and salts are determined as directed for
blood. The lactose can be obtained as directed by
Pavy's method for separating glucose from blood, or
can be estimated directly by Fehling's process (*see*
Urine), after the milk has been freed from albumin.

CHAPTER IV.

MORBID CONDITIONS OF URINE.

105. **Examination of morbid urine.**— In
order to draw conclusive results from the examination
of urine in disease, a systematic plan of procedure must
be adopted and the circumstances and conditions
under which the secretion was passed precisely stated.

The neglect of these precautions renders many investigations valueless, whilst in some instances it leads to erroneous conclusions being drawn. As is well known, the specific gravity, the reaction, and proportionate relationship of the different constituents of the urine, even in health, vary considerably during different periods of the day, whilst in disease such variations are still more considerable. It is evident, therefore, if we wish to arrive at a right conclusion as to the nature of the pathological condition with which we have to deal, our observations must be based upon an examination of the whole of the urine passed during a period of twenty-four hours. If that is not always possible, from at least two samples; the one taken on first rising in the morning, the other two hours after the principal meal of the day. *No definite conclusion should ever be drawn from the examination of a single sample of urine passed within twenty hours.* In noting the character and qualities of urine in disease the following scheme may advantageously be followed : (1) State the circumstances under which the urine was passed; whether it represents the secretion of the twenty-four hours, or is only a specimen passed at some period during the day; if the latter, state the time when voided. (2) Record the quantity presented for examination (if the twenty-four hours' urine has been collected, give the amount), the specific gravity and reaction, note also its colour and odour and degree of clearness. (3) Test for abnormal products, sugar, albumin, etc. (4) Collect deposit for chemical and microscopic examination. This procedure will give us an insight into the *qualitative changes* that occur from day to day in the character of urine during the progress of disease. To determine the *quantitative changes* in the amount of the urinary constituents two methods are employed : (1) *Exact chemical determination*, which consists either in precipitating, collecting, and then weighing the precipitated substance (the "gravimetric method"); or by precipitating or otherwise altering the substance by means of a solution of a reagent of known strength, or ascertaining the quantity of the reagent required to effect a complete change; this is the "volumetric method." (2) *Approximate estimation* by some ready means and easily applied clinical method, such as calculating the amount of urinary solids from the specific gravity, or the quantity of sugar by the loss of weight occasioned by fermentation with yeast, or by the degree of intensity of colour when compared with a standard of comparison. Approximate estimation, however useful, should never be entirely relied on. Exact chemical determinations should always be made from time to time. In all cases the calculation should be made from the urine passed in a period of twenty-four hours; in this way we get the absolute as well as the relative amount of the substances passing out of the body during this period.

106. **Variation in the urinary water and solids.— Specific gravity.**—The following table gives approximately the amount of the chief constituents of the twenty-four hours' urine, in a child, a growing lad, and an adult.

MEAN AMOUNT CHIEF CONSTITUENTS OF NORMAL HUMAN URINE PASSED IN TWENTY-FOUR HOURS.

	5 years weight 37½ lbs.	12 years weight 64½ lbs.	35 years weight 147 lbs.
Water . . .	450 cc.	860 cc.	1,450 cc.
Urea . . .	11·1 grms.	16·7 grms.	33·4 grms.
Uric acid . .	0·4 ,,	0·6 ,,	0·8 ,,
Alkaline phosphates	0·9 ,,	1·6 ,,	2·1 ,,
Earthy phosphates .	0·5 ,.	0·8 ,,	1·3 ,,
Chlorides . .	1·2 ,,	3·4 ,,	6·2 ,,
Sulphates . .	0·7 ,,	1·4 ,,	2·6 ,,

In calculating the amount of solids passed in disease, we must allow a range of one-fourth of the mean amount, above and below that amount. The quantity of urine passed in the twenty-four hours is, however, no measure of the amount of solid matter passing out of the body by the kidneys, since 30 ounces of urine may contain as much solid matter as 60 ounces. The amount of solid matter is therefore determined either by evaporating the urine and weighing the residue, or else by taking the specific gravity of the urine by means of a urinometer, and calculating the solids from the density. The former process is tedious, and requires an expenditure of time which precludes its employment when examinations have to be made daily. For clinical purposes the solids may be calculated, when precautions are taken to ensure accuracy, from the specific gravity. The calculation is based on the fact that normal human urine of twenty-four hours contains 4 per cent. of solid matter, and that the specific gravity of that urine is registered at about 1·020. By multiplying the two last figures of the specific gravity by 2 we get 40 in every 1,000 parts, or exactly 4 per cent. A thousand grains, therefore, of urine, specific gravity 1·020, contains 40 grains of solid matter; or, if French measures are employed, 1000 cubic centimetres of the same specific gravity contain 40 grammes. As stated in the above table, the quantity of water passed in twenty-four hours is 1450 centimetres, or 50 ounces; then the quantity of solids as calculated from the specific gravity discharged in the same period will amount to 2 ounces, or 54 grms. In making calculations based on the specific gravity, it is important to note the temperature at the time the observation is made, since a difference of 7° F. from the temperature at which the urinometer was graduated represents a difference of one degree. The observation of the specific gravity in connection with the amount of urine passed in the twenty-four hours, affords in itself an important indication as to the metamorphoses going on in the body.

Under the terms hydruria, diabetes, polyuria, etc., authors have described certain morbid conditions of the urine, characterised by excessive and persistent discharge. Most authors apply either of the above terms indifferently, without reference to the quantitative relationship that may exist between the urinary water and solids. The following classification will assist the memory :

(1) *Hydruria.*—A copious discharge of aqueous urine. In these cases there may be a decrease in the solid constituents of the urine, but there is certainly no increase. Extreme instances of hydruria are met with in cases of "diabetes insipidus," in which the daily urinary flow may amount to 7,000 to 9,000 cubic centimetres of a specific gravity of 1·002. As a temporary condition it is frequently met with in hysterical females and persons of highly neurotic temperament. Associated with minute traces of albumin, it is a condition of urine met with in granular kidney ; the amount of diuresis, however, is not so extreme, nor is the specific gravity so low, as in typical instances of diabetes insipidus.

(2) *Polyuria.*—An increase of water with an increase of urinary solids, dependent on increased tissue metabolism. In these cases all the urinary constituents seem to be increased. In this division we have those cases of increased elimination of urea to which Prout assigned the term "azoturia"; and those cases which Dr. Tessier has described as "phosphatic diabetes," where the amount of phosphoric acid excreted is enormously increased. "Azoturia" and "phosphatic diabetes" are probably allied

conditions, due to increased tissue metabolism taking place under nervous disturbance. Both forms have been met with associated with the following conditions : (*a*) Cases in which nervous symptoms are predominant; (*b*) accompanying pulmonary consumption ; (*c*) cases which alternate or co-exist with saccharine diabetes; (*d*) which run a distinct course, resembling saccharine diabetes, but without sugar.

(3) *Diabetes mellitus.*—Increase of the urinary water, together with constant (unless checked by diet) excretion of glucose in excessive amounts. The occasional appearance of sugar in urine is not to be taken as an indication of true diabetes mellitus, but as due to temporary functional disturbance of the liver, though it must not be overlooked that temporary "glucosuria" is often precursory of the more serious disease. In cases of diabetes mellitus the relationship between the amount of urine excreted and the urinary solids are very various. In typical cases the relationship is pretty constant; whilst the water is considerably increased there is superabundance of sugar, and the urea, probably owing to increase of nitrogenous diet, is considerably in excess. In a second class of cases we find a considerable increase in the amount of urine, the sugar moderate in quantity, the urea not much increased, in some cases even below the normal, and the urine frequently albuminous. In a third group of cases the urinary water is only moderately increased, the sugar rarely exceeding $2\frac{1}{2}$ per cent., whilst the urea is generally considerably in excess, more than can be accounted for by increase of nitrogen ingested. No satisfactory explanation has been offered to account for these variations, the latter, as the least frequently observed, is probably an early stage of the former ; the marked increase of the urea excreted, as compared to the

moderate amount of sugar generally noticeable, points
to increased tissue metabolism. These cases, after a
comparatively mild course, often suddenly develop
into acute diabetes, and run a rapid course, ending in
diabetic coma.

Baruria.—The urinary water is not increased, may
even be diminished, but the urinary solids are in excess.
From the concentrated urine, urates are frequently
deposited. Dr. Fuller was the first to apply the term
baruria (βαρύς, heavy) to this class of urines, which he
associated with certain forms of dyspepsia (*Med.-Chir.
Trans.*, vol. li.). It is a condition of urine found in
a class of cases described by Murchison as due to
lithæmia.

Anazoturia.—This term was originally applied by
Willis to a class of cases in which there was a copious
discharge of pallid urine, with marked deficiency of
urea. These cases, however, should be referred, I
think, to the class hydruria. The term anazoturia is
better applied to the cases described by Dr. Andrew
Clarke as due to "renal inadequacy." In these there
is no marked increase in the amount of urine passed,
but a notable deficiency in the amount of urea
excreted. As it is still uncertain whether this condition depends on inadequacy of the renal functions,
or on deficient metamorphosis of tissue generally ; a
term which defines the actual condition of the urine,
rather than one which commits us to an hypothesis, is
for the present decidedly the safest.

107. **Reaction.**—Within a period of twenty-
four hours the reaction of healthy urine will be found
to vary considerably. Thus before meals it will have
a high range of acidity, whilst after food it will become
nearly neutral, or even sometimes alkaline. This
depression of acidity, which has been called the alkaline tide, has been accounted for by Dr. Bence Jones,
by the fact that at the moment of its occurrence,

acid is being drawn from the circulation to supply
the gastric juice. Dr. Roberts, of Manchester, however, regards the depression as due to the introduction of newly-digested food into the blood, which
supplies it with alkaline bases. Although both explanations are based upon actual observation, yet as
the acidity of the urine can be depressed, and even
rendered alkaline, by other circumstances besides those
attendant on digestion, as the mere act of rising in the
morning before breakfast is taken, or the cold douche,
or sweating in the vapour bath, some other explanation
must be offered to explain the depression in these
cases. And this, I think,* is to be found in the fact
that there is another channel by which acid is withdrawn from the blood beside the gastric secretion, and
that is by the lungs. In the explanations hitherto
advanced to account for the phenomenon of the
alkaline tide in the urine, this fact has not received
attention. Dr. Edward Smith, in his researches "On
the Elimination of Carbonic Acid," has shown conclusively that the exhalation of carbonic acid by the
lungs is increased by food and diminished by fasting,
and that the amount exhaled during sleep is considerably less than is set free in the waking state. It
therefore happens that the time when most carbonic
acid is being exhaled corresponds with the time when
observers have noticed a decided diminution in the
acidity of the urine, whilst the circumstances that
diminish the exhalation of carbonic acid (namely,
sleep and fasting), are attended by a rise in the acidity
of the urinary secretion.

The acid reaction of the urine is chiefly, if not
entirely, due to the presence of acid sodium phosphate, and occasionally to an excess of acid salts of
hippuric and uric acids. It is only recently that an

* " Morbid Conditions of Urine dependent on Derangements of
Digestion." Churchill, 1882.

explanation has been offered to account for the seeming paradox of the separation of an acid secretion like urine from alkaline blood. In 1874 I pointed out* that it might be the result of the decomposition between the neutral sodium phosphate and acid sodium carbonate (bicarbonate), both of which exist in the blood, resulting in the formation of acid sodium phosphate, and normal sodium carbonate. The former diffuses out through the renal parenchyma, whilst the latter remains in the blood (§ 93). Maly, however, believes that acid sodium phosphate exists in a free state in the blood, and he has shown that if a mixture of neutral sodium phosphate and acid sodium phosphate be placed together in a dialyser, the acid salt passes into the surrounding distilled water. Maly's explanation has the merit of simplicity, but it does not wholly account for many of the phenomena connected with the variations in the reaction of the urine.   If, on the other hand, the view that the acidity of the urine is caused by the reaction between acid sodium carbonate and neutral sodium phosphate be accepted, it will explain another paradox which has been observed by Bence Jones, Beneke, Parkes, and myself, to occur after the administration of the bicarbonates (acid carbonates) of ammonia, potash, and soda, under certain conditions, viz. causing of an increased acidity of the urine.   The free acidity of the urine is reckoned as oxalic acid, and in the healthy state the total acidity of the twenty-four hours' urine is equal to a degree of acidity represented by from 1·5 grms. to 2 grms. of oxalic acid.

After it has been passed, the urine, if originally acid, increases its degree of acidity owing to the acid fermentation of the pigment and extractive matters ; the highest degree of this acid fermentation is reached about the third day.   The urine then

* *Lancet,* July 21, 1874.

gradually becomes alkaline from the ureal decomposition.

In disease the acidity of the urine may become persistently highly acid or persistently alkaline (fixed or volatile), or there may be fluctuations, a high degree of acidity alternating with a neutral or alkaline condition. The variations in the reaction of the urine in disease will therefore be considered under three heads.

(1) *Highly acid urine.*—Urine may become more acid, relatively, owing to concentration of the urine. Thus, in hot weather, owing to the increased action of the skin, the amount of urinary water is lessened and the urine becomes denser. Similarly in pyrexia, especially if attended with profuse sweating, as in rheumatic fever and in diarrhœa. In diabetes mellitus the acidity of the urine is considerably increased ; when freshly passed it has a degree of acidity considerably above the average, and becomes more acid after being kept some hours, owing to lactic and acetic acid fermentation which takes place. In acid dyspepsia, so called on account of its supposed association with hypersecretion of acid gastric juice, the urine is at times highly acid, alternately, however, with urine that is neutral or even alkaline. This variability in the reaction of the urine is frequently to be met with in children, in whom irregular secretion of the gastric juice is very readily excited. The degree of acidity is determined by neutralising 100 cc. of urine with a solution of sodium hydrate, standardised so that 1 cc. = ·01 grm. of oxalic acid. The number of cc.'s of the standard solution employed to effect neutralisation, multiplied by ·01 gives the degree of acidity in 100 cc. of urine, and from this the total acidity of the twenty-four hours' urine can be calculated.

(2) *Urine alkaline from fixed alkali* —This condition is due either to excess of the alkaline carbonates

of soda and potash, or the alkaline phosphates; often to
an excess of both. (*a*) Excess of alkaline carbonates :
The urine effervesces on the addition of strong acids ;
it is generally turbid from precipitation of the earthy
phosphates, though these are not necessarily excreted
in excess. Indeed, in many cases I have found them
diminished ; on the other hand, the urine may contain
an excess of uric acid. This condition of urine may
arise, (1) from general debility and the feebleness with
which the respiratory act is performed, leading to the
accumulation of carbonic acid in the system. With
regard to this point, it is interesting to note that
urine alkaline, from the presence of carbonates of
the fixed alkalies, is frequently met with in patients
convalescing from acute diseases. (2) Diminished
secretion of bile, which is the frequent result of the
duodenal catarrh produced by the irritation of the acid
contents of the stomach being poured into the intestines,
gives rise to an accumulation of alkaline carbonates in
the blood, the bile being the chief secretion by which
alkaline salts are removed from the body ; for though
a portion of them are undoubtedly reabsorbed into the
blood from the intestines, a considerable proportion
of them are discharged with the fæces. Obstruction,
therefore, to the discharge of bile leads to their retention
in the blood, and consequently being eliminated in
greater quantity by the kidney. (3) The acids formed
by fermentative changes being of the fatty acid series,
these, on entering the system, are oxydised into carbonic
acid, and this uniting with the bases of the alkaline
oxides, forms carbonates of these bodies, and by increasing the alkalescence of the blood will diminish the
natural acidity of the urine and even render it alkaline.
The dyspepsia generally associated with this form of
alkaline urine is attended with great depression of
spirits, the bowels are constipated, flatulence is a prominent symptom, the skin is sallow and dry, and the

I

functions of the liver evidently deranged. The urine, after remaining alkaline for some days, depositing oxalates and phosphates, often becomes suddenly acid, and deposits large quantities of uric acid.

(b) Excess of alkaline phosphates of soda and potash : Little is known regarding the pathological significance of urine alkaline from this cause. It is, however, frequently met with in neurotic individuals. In the majority of cases the earthy phosphates are increased as well as the alkaline carbonates. Excessive elimination of the alkaline phosphates has been noticed in cases of acute inflammation of the membranes of the brain, in the acute paroxysms of certain forms of mania, after injuries to the head, and in certain obscure spinal affections, probably functional in character. The urine, in these cases, is generally increased in quantity ; on passing from the bladder the first portion may be clear and the remainder thick. Sometimes a small quantity, consisting almost entirely of amorphous phosphate of lime, may be passed with considerable straining and feeling of irritation immediately after the flow. Urine alkaline, from either excess of alkaline carbonates or phosphates, must not be confounded with true phosphaturia when there is an excessive excretion of earthy phosphates. In simply alkaline urines there may be a deposit of earthy phosphates, though there need not necessarily be an excess. In phosphaturia, even with considerable excess of calcium phosphate, there is no deposit, and the urine is often acid (§ 113). The degree of alkalescence is determined by neutralising 100 cc. of the urine by a solution of oxalic acid, standardised so that 1 cc. = ·01 grm. of sodium hydrate ; the number of cc.'s of the standard solution employed to effect neutralisation multiplied by ·01 gives the degree of alkalescence in 100 parts ; and from this the total alkalinity of the twenty-four hours' urine can be calculated.

(3) *Urine alkaline from volatile alkali* (ammonium carbonate).—This condition is induced by disease of the genito-urinary organs, since experiments on healthy animals show that the urine does not become ammoniacal by prolonged retention in the bladder, so long as that organ does not become inflamed, and also that the introduction of the "ammoniacal ferment" into the bladder of animals will not cause decomposition of urea so long as the mucous membrane remains healthy. Feltz and Ritter, from observations made on seventy-eight persons suffering from different diseases, came to the conclusion that the urine does not become ammoniacal unless received into dirty vessels, or mixed with products of the decomposition from the mucous surfaces of the genito-urinary apparatus. In one of their cases a patient was taking 0·2 gramme of bichloride of mercury daily, and the urine was analysed every day ; the acidity fell on the ninth day to ·02 gramme, on the tenth day it was neutral, on the eleventh day it was alkaline and ammoniacal. This change in the reaction coincided with the appearance of a trace of albumin in the urine, which was turbid, and contained flakes of epithelium and leucocytes. In some cases of scarlet fever I have noticed that when albumin appeared in the urine the reaction frequently became alkaline, and crystals of ammonio-magnesium phosphates were deposited. Dr. Owen Rees has advanced a theory that it is by no means necessary for ammoniacal urine to depend on decomposition of the urea ; he maintains it can be formed by the secretion of the mucous membrane, which owes its alkalinity to fixed alkali, and which, mixed or mingled with the urine, unites with the acids of the ammoniacal salts and thus liberates the ammonia. In answer to this view it is sufficient to state that the existence of ammoniacal salts in the urine, except as the result of the decomposition of urea, has been

denied by most chemists, and that if Dr. Owen Rees' view were correct, ammoniacal urine would be more frequent than it is, since whenever the mucous secretion of the urinary passages was increased the urine would become ammoniacal. Clinical experience teaches us that this is not the case. At the same time, there can be no doubt that the presence of fixed alkali in urine greatly favours ureal decomposition, and the process is induced more rapidly. Whenever the urine becomes ammoniacal, crystals of ammonio-magnesium phosphates are formed, which either pass away as gravel or are retained as a calculous deposit. We distinguish between the reaction due to fixed alkali and that of volatile alkali by the fact that with the former the blue colour given to red litmus does not disappear on drying, whilst with the alkalescence due to ammonia the blue tint is evanescent.

108. **Colour.**—The nature of the pigment that imparts the colour to the urine has been the subject of much discussion. Spectroscopic analysis has recently thrown fresh light on the subject. MacMunn has shown that human urine always gives an absorption band at F in the same manner as *choletelin*, the pigment obtained by Jaffé from bile. He thinks all the colouring matters of the bile are produced from hæmatin by reduction, due to the action of the bile acids on hæmoglobin. All the colouring matter of the bile, including hæmatin, urobilin of biliary origin, bilirubin, etc., are oxydised to *choletelin*, and there is evidence to show that blood serum contains this body on its way to be excreted by the kidneys. The urobilin of the bile is produced in the intestine, and may, in certain conditions of the system, appear as such in the urine, but under normal conditions is oxydised into *choletelin*, which must be considered one of the chief urinary pigments. Most of the urinary pigments may

thus be traced back to urobilin of biliary origin; but
there is also evidence to show that some of them are
derived from hæmatin directly, and that pigments
derived from that source may occasionally entirely
replace the normal pigment. The spectroscopic characters of the pigment differ. In febrile urine the black
band in F is sharp, in normal urine it is less marked
at the edges and less shaded. The band is well seen
in alcoholic solutions, and is destroyed by the action
of caustic alkalies. In addition to the pigment above
described as related to urobilin and convertible into
it, most urines contain a pigment allied to, if not
identical with, indican, and which is commonly known
as uroxanthin. The source of this body is probably
from indol formed by the decomposition of proteid
substances by pancreatic digestion, since indican is
found in the urine of animals after the subcutaneous
injection of indol, and also after ligature of the small
intestines (§ 76). In many diseases the amount
that appears in the urine is considerably increased,
as in diabetes, dysentery, the reaction stage of
cholera, in obstruction, and other affections of the
intestines. Urines containing this substance, when
heated with strong acids, give rise to blue, greenish,
and red pigments (the blue and green urines sometimes met with in disease may thus be referred to
indican in different states of oxydation); these sometimes occur spontaneously in urine. Urines containing much indican are invariably highly acid (and
this acidity increases on keeping), and they generally
deposit a considerable amount of uric acid, and
partially decompose a solution of sulphate of copper.
Whether this reduction is due to the presence of uric
acid in excess, or to the fact that indican is a glucoside, and yields on decomposition a molecule of
glucose, is not determined. *Melanin*, a black pigment,
occurs pathologically in the urine in patients suffering

from melanotic tumours; it is sometimes found in the
urine of persons suffering from ague. It is soluble in
caustic potash, and can then be decolorised by passing
chlorine through the solution. The colour of the
twenty-four hours' urine is light amber, the urina
sanguinis and urina cibi a golden amber, the urina
potu a pale straw colour. It must not be forgotten
that many articles of diet and medicine impart a
colour to the urine. The presence of blood, bile,
albumin, sugar, change the colour of the urine.
These alterations will be considered under their
respective heads.

109. **Odour.**—Normal urine has an odour *sui
generis*. It is described as aromatic. Alkaline urine
evolves an ammoniacal odour when its alkalinity is
due to volatile alkali; a faint mawkish smell, like that
of horses' urine, when alkaline from fixed alkali.
Diabetic urine is said to exhale a whey-like fragrance.
Urine containing cystin at first smells like sweet-
briar, but speedily becomes horribly offensive. In
certain forms of dyspepsia the urine has a sickly
penetrating odour. Medicines and certain articles of
food often impart a peculiar odour to urine, as turpentine, the fragrance of violets, and asparagus, a
peculiarly rank foxy odour.

110. **Urea** CH_4N_2O.—About one ounce of this
body, or, if we calculate in French measures, about
33 grammes of this substance, are passed out of
the body in the urine in the course of twenty-four
hours. We have already stated (§ 79) that urea is
isomeric with ammonium cyanate. With regard to this,
attention has been called to the fact that urea is a much
more stable body than ammonium cyanate, and that
in the transformation of the latter into the former,
energy is set free. Thus, ammonium cyanate is the
type of living, and urea of effete, nitrogen, and the
conversion of the former into the latter is the image

of the essential change which takes place when a living
proteid dies. It is probable, moreover, that cyanogen
compounds precede the formation of urea, and act
with great molecular energy till they pass into the
more stable but effete form of urea, when they are
cast out of the body. Urea thus represents the ultimate product of the metabolism of the nitrogenous
constituents of the food and tissues. In health, the
amount excreted is proportionate to this metabolism;
in disease, however, no such relationship is maintained.
The amount is increased in all acute diseases, and is
especially marked during pyrexial exacerbations. In
typhus fever the excretion is highest during the first
week, the excretion then being often double that of
the fourth week, although the patient during the first
stage is on low diet, and during the latter period on
meat diet. In relapsing fever the urea is increased
during the paroxysms, and diminished during the in-
terval. In enteric fever, urea is excreted in the largest
amounts during the first week of the disease, it then
gradually diminishes. Still the quantity continues in
excess of the normal standard as long as the fever
lasts; the amount excreted daily during the disease is
not influenced by the amount of diarrhœa. In eruptive fevers as measles, small-pox, and scarlet fever,
urea is increased in amount during the first four days.
If, in the latter disease, kidney complication sets in,
a sudden fall takes place. In intermittent fevers,
the urea of the twenty-four hours is not markedly
increased, but during a paroxysm, and just before it,
there is a decided increase, followed by a decrease.
The reason of the increase of urea accompanying or
preceding a rise of temperature, has been given (§ 7).
As the liver is the organ in which metabolic changes
leading to the formation of urea are the most active,
though it has now been shown most conclusively that
urea is also produced largely from the leucin, the result

of pancreatic digestion, and from the kreatin in muscles, disease or disturbance of functions of that organ especially affect the excretion. In acute yellow atrophy the urea is slightly increased at first, during the hyper-æmic stage, but rapidly decreases as the disease advances, and when the liver cells are destroyed, nearly all trace may disappear from the urine, its place being taken by uric acid, leucin, and tyrosin, and some ill-defined albuminoid bodies resembling peptones. In hepatic abscess and in cancer of liver a notable diminution is observable. In diseases of the kidney, in both acute and chronic forms of nephritis, the excretion of urea is lessened; although no relationship exists between the discharge of albumin and excretion of urea, still a more favourable prognosis may be expressed when the urea does not progressively diminish. In the albuminuria of pregnant women the amount of urea in the urine is only slightly diminished. In a case of Dr. Maxwell's, of Woolwich, I found urea in the twenty-four hours' urine to the amount of 27 grammes, a little below the normal. In a case of Dr. John Williams', in the urine of twelve hours immediately preceding venesection the urea was 14·26 grammes, and the albumin was 4·9 grammes; in the twelve hours' urine immediately succeeding venesection, I found urea 15·6 grammes, and the albumin 2·6 grammes. Bleeding, therefore, had little effect on the excretion of the urea, which in both periods was only little below the normal, but it materially reduced the amount of albumin. In diabetes, urea is always largely in excess, in measure due no doubt to the increased quantity of animal food consumed; still this does not altogether explain the whole of the increase. A sudden fall in the excretion is an unfavourable sign, as it often is a prelude to diabetic coma, in which state both urea and sugar are excreted in lessened quantities; the fall in the urea usually

precedes that of the sugar. In uterine diseases a
temporary excess of urea in the urine is to be met
with. It is said to be increased both before and after,
but diminished during, the menstrual period. In the
periodic jaundice recurring at the menstrual periods
(*icterus menstrualis*) urea is increased. In phthisis
the excretion of urea corresponds with the
pyrexia. Rapidly growing cancer causes a
diminution, a fall from 29·5 grammes to as
low as 6·9 has been recorded. In certain
constitutional states urea is also diminished,
as in patients suffering from anaemia, hydraemia,
chronic alcoholism, and syphilitic cachexia.

Fig. 4.—
Pure
Urea.

Under the term "renal inadequacy," Dr.
Andrew Clarke has described condition in which only
a very small quantity of urea is continuously passed,
although there is no recognisable disease present.
These cases improve on a carefully regulated dietary. As a converse to these are
cases which habitually excrete an enormous
amount of urea, together with the other
urinary constituents. Dr. Prout termed
this condition azoturia, though it may
more appropriately be termed polyuria; it
often precludes phthisis, or may be an

Fig. 5.—Ni-
trate of
Urea.

antecedent condition of diabetes mellites. (For
qualitative tests for urea, see § 7.) For clinical
purposes urea is often roughly estimated by the
precipitation of nitrate of urea, by adding an equal
volume of strong nitric acid to the urine; if nitrate
of urea is thrown down without concentration of the
urine, then it is said to be in excess, if concentrated
to half its bulk, about normal; if further concentration is required, then less than normal. It is needless
to point out the fallacy of this method, since unless
the quantity of urine passed and the specific gravity
are likewise noted, urines containing absolutely a

considerable quantity of urea may be passed by the patient in a very dilute form, whilst others, containing only an ordinary quantity, may be voided in a concentrated state, and give the reaction, and thus mislead the physician. For accurate determination, recourse must be had to volumetric analysis: (*a*) *Liebig's method.* For this purpose 40 cc. of urine are taken, and freed from albumin if present by heat and filtration; and mixed with exactly the same quantity of a solution of barium hydrate (2 volumes) with barium nitrate (1 volume); this precipitates the phosphates and sulphates; a few drops of solution of nitrate of silver are then added, which precipitates the chlorides. Set aside till the precipitate has collected at the bottom of the beaker, then filter, and of the clear filtered solution take 20 cc., reserving the remainder in case of accident. This of course represents 10 cc. of urine. Now run into this solution, from a burette, 10 cc. of standardised solution of mercuric nitrate, stir the mixture well, withdraw a drop on a glass rod, and let it fall on a drop of sodium carbonate solution placed on a white plate or on a flat porcelain dish. If a yellow stain occurs the process must be repeated with the reserve stock; but this is unlikely, unless the urea is very inconsiderable indeed. If there is no stain, add 5 cc. more, and then test again; if no reaction occurs, repeat the process more cautiously, adding 1 cc. at a time till a yellow stain, due to the formation of hydrated oxide of mercury, is at last produced. Now as each 1 cc. of the standard solution of mercuric nitrate is equivalent to ·01 gramme of urea, the number of cc.'s employed to produce the yellow stain indicates the number of grms. present in 10 cc. of urine, from which the amount in the twenty-four hours' urine can readily be deduced. (*b*) *Russell and West's method* is based on the fact that hypobromous acid decomposes urea into water, carbonic acid, or nitrogen. The latter gas is collected alone in a graduated tube, which is standardised so that each measure represents one gramme of urea in 100 cc. of urine. In employing this test for the determination of urea in diabetic urines, it must be remembered that grape sugar increases the quantity of nitrogen evolved from urea by sodium hypobromite by quite seven per cent. The deficiency of nitrogen yielded with pure solution of urea, under the hyperbromite test, is about eight per cent., the addition of glucose, therefore, brings it up to the theoretic yield. This is of little importance unless the analyses are made for purpose of comparison of diabetic with non-saccharine urine. In making a series of observations, care must be taken always to secure the same temperature, as a slight decrease or increase makes a considerable difference in the gas volume. From non-attention to this important particular, many discordant results have been obtained. After considerable experience I have come to the conclusion that Liebig's method is not only the most reliable, but after a little experience quite as readily performed as the other.

111. **Uric acid** $C_5H_4N_4O_3$.—Was discovered by Scheele in 1776, and was at first thought to be solely a constituent of urinary calculi, hence the term *lithic acid* usually applied to it. In 1797, Woollaston showed that gouty tophi were composed of sodium urate; whilst, in 1848, Dr. Garrod brought forward the fact that in true gout an excess of uric acid exists in the blood prior to and at the period of the attack.

From the circumstance that uric acid is a di-ureide, that is, by oxydation a molecule of uric acid can be split up into a molecule of a non-nitrogenous acid and two molecules of urea, it has been assumed that when the process of oxydation is imperfectly performed within the body uric acid will be found in excess in the blood ; and this assumption has been further strengthened by the supposition that uric acid is one of the substances through which each particle of albumin passes before it is thrown out of the body. This view has for many years completely dominated urinary pathology. Now, however, since it has been shown that uric acid is not a necessary antecedent of urea, which is largely formed from kreatin in muscle, and leucin and other bodies in the alimentary canal, the view has gained ground that uric acid in the human body in health is only formed in minute quantities, and that even in disease it is not formed in anything like the amount formerly supposed, and that when it is deposited from the urine or in the tissues, the fact of the occurrence of such deposit may be generally referred to its insolubility rather than excessive production in the system. It is now taught that while uric acid is met with in small quantities in the large glands like the spleen, liver, etc., in health it is never found in the blood ; so that uric acid is probably oxydised as soon as formed, and that the small quantity found in normal urine, only 0·5 gramme, or about 7 grains, a whole day's excretion, is not derived from the blood, but from the kidney, which, instead of being oxydised, as is the case with other organs, passes away with the secreted urine. In many diseases attended with considerable tissue metamorphosis, the amount of uric acid formed in the large organs is increased, and a portion not completely oxydised passes into the blood, and from thence into the urine, though even then the amount is never large, rarely exceeding 1·5 grammes in the twenty-four hours as an outside average ; in these cases there is, as well, generally an increase in the amount of urea excreted. In gout there may be an increased production, but most likely, as Dr. Garrod suggests, it is due rather to an accumulation in the blood, caused by a resorption in the uric acid formed in the kidney not being excreted

with the urine, but taken up by the blood and carried
the round of the circulation in combination with soda,
and deposited in the least vascular parts, the cartilages
of the joints, the cartilages of the ear, the straight
tubules of the kidney, etc., as sodium urate. Uric
acid is the most insoluble of all the substances formed
in the body, requiring 15,000 parts of water for
solution, whilst urea is soluble in its own weight. It
is, therefore, fortunate that those animals whose
urinary apparatus is not adapted for carrying off solid
or semi-solid urine like birds and reptiles, that soluble
urea replaces the insoluble uric acid, otherwise
calculous disease would be infinitely more common
than it is. Owing to this insolubility, whenever the
amount of water in the urine required to keep uric acid
and its salts in solution falls below a certain point,
then uric acid or its salts are deposited. In acid conditions of the urine, uric acid and its salts, unless
heated, are almost altogether insoluble, so that when
the natural acidity of the urine is at all heightened,
they are at once deposited. The following summary
gives the chief conditions which lead to a deposit of
uric acid in the urine, or its excessive elimination
from the body.

(A) Deposits of uric acid or urates, not, however,
necessarily eliminated in excessive quantities.

(1) *Absolute increase in the acidity of the urine.*—
The occasional deposit of urates observed in winter
arises from this cause. The action of the skin being
checked, the acidity of the urine increases during cold
weather. Similarly in many extensive cutaneous
diseases, such as eczema and psoriasis, uric acid
deposits are of frequent occurrence ; also in forms of
dyspepsia associated with irregular secretion of gastric
juice.

(2) *Relative increase in the acidity of the urine.*—
The deposits of urates frequently noticed during the

summer months originate in this way; the cutaneous transpiration being increased in hot weather, the urine is more concentrated. Similarly in pyrexia, especially rheumatic fever, and in diarrhœa. Uric acid deposits alternating with sugar are often caused in this way; since as the sugar disappears urination is not so profuse, and a relative increase in the acidity of the urine occurs. This relative increase

Fig. 6.—Forms of Uric Acid.

may not only be caused by a diminution of the water excreted, but from deficiency of the alkaline phosphates; this condition is frequently met with in the urines of ill-nourished or strumous children.

(B) Uric acid eliminated in excess, but not necessarily deposited from the urine.

(1) *Uric acid in excess usually attended with a diminution of the other urinary constituents (true lithæmia.)*—Chiefly in diseases of the liver, such as acute yellow atrophy, cirrhosis, and cancer. In diseases of the spleen, leucocythæmia. In scurvy an excess of uric acid is generally observed, with a diminution of urea and the alkaline phosphates.

(2) *Uric acid in excess attended with an increase of the other urinary constituents.*—In functional derangements of the liver, especially those brought about by disturbance of the "nitrogenous equilibrium" by the ingestion of too much animal food. As a condition antecedent to the development of phthisis or cancer, and sometimes of diabetes, or preceding the outbreak of such constitutional conditions as syphilis, scrofula, and of gout in its early attacks.

Uric acid when deposited from the urine in a free state resembles grains of cayenne pepper, and presents

under the microscope the appearance shown in
Fig. 6. The bases met with in urine in combination with uric acid are chiefly those of soda
and ammonia, though lime is sometimes present.
They are thrown down as a granular deposit, but
amidst the small granules are semi-crystalline bodies
depicted (Fig. 7). Both uric acid and its salts
are freely soluble in alkaline solutions; this is an
important point to remember, not only
as regards treatment, but as regards
their detection in the urine, since their
crystals are frequently modified in
shape, may be mistaken for other
urinary deposits; the fact of their
dissolving in liquor potassæ at once
identifies them. Uric acid and urates
yield a magnificent purple when heated
with nitric acid, and the dry residue

Fig. 7.— Forms
of Urate of
Sodium (a);
Urate of Am-
monia (b).

is touched with ammonia (§ 64). When a concentrated solution of urates is treated with strong
nitric acid, the uric acid is liberated in an amorphous
form, in which it is more soluble than in its crystalline
state; on standing it slowly recombines with the
sodium salts in the urine, and is deposited as urate of
sodium. The peculiar bulky gelatinous precipitate,
soluble when heated, some concentrated urines give
when tested with nitric acid for albumin, is due to the
uric acid being liberated in this amorphous form.

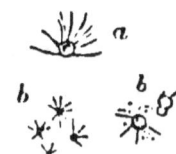

Uric acid is determined quantitatively by precipitating the uric acid from the urine, and by collecting
and weighing the crystals. For this purpose place 100
cubic centimetres of filtered urine (having previously
dissolved any deposited urates by heating) in a glass
vessel, and add 10 cubic centimetres of strong hydrochloric acid; set aside in a cool dark place for twenty-
four hours. Then carefully collect crystals, and transfer
them to a watch-glass, wash thoroughly with dilute

hydrochloric acid, and then transfer them to a weighed filter. Dry in hot-air bath and weigh. The increase in weight over that of the filter will give the amount of uric acid in 100 cubic centimetres of urine.

If the urine is of low specific gravity, below 1·015, it should be concentrated to one-third its bulk, if below 1·010 to one-half.

112. **Oxalate of lime** CaC_2O_4.—An extremely small quantity of oxalic acid is met with in all urines, chiefly in combination with ammonia and soda, forming soluble salts. In several morbid states of the system, however, oxalic acid appears in the urine as oxalate of lime, in which case it either comes away as a fine crystalline deposit, or else is retained in the urinary passages to lay the foundation of a calculus (mulberry calculus). Few subjects in urinary patho-logy have excited keener controversy than that relating to the causes tending to produce deposits of oxalate of lime crystals in urine. The view has been generally held that oxalate of lime in urine was simply derived from the decomposition of uric acid after it had been passed, and that the presence of oxalate of lime in the urine meant nothing more than increased excretion of uric acid. This view was based on the assumption that oxalic acid represented the imperfect oxydation of uric acid. It will be seen, however, from the following table, that so far from oxalic acid being a product of imperfect oxydation of uric acid, it is only obtainable by oxydation being carried to its ultimate stage : thus

Uric acid.   Alloxan.   Urea.

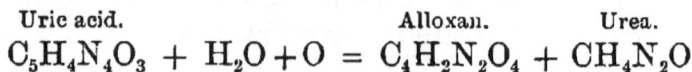

$$C_5H_4N_4O_3 + H_2O + O = C_4H_2N_2O_4 + CH_4N_2O$$

and

Alloxan.   Mesoxalic acid.   Urea.

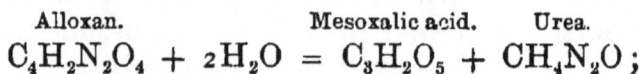

$$C_4H_2N_2O_4 + 2H_2O = C_3H_2O_5 + CH_4N_2O\,;$$

and mesoxalic acid by further oxydation yields carbonic acid and oxalic acid. It is, therefore, manifestly incorrect to speak, as most writers on urinary pathology have done, of oxalic acid as the imperfectly oxydised product of uric acid ; on the contrary, oxalic acid is only obtained from uric acid by oxydation being carried to its ultimate stage. Now within the body, under the influence of increased oxydation, this reduction of uric acid may occur ; but as a considerable quantity of oxygen is required to effect the reduction, and as the clinical and pathological conditions in which we meet with oxalate of lime in the urine, do not point to increased oxydation going on within the body, but the reverse, it is manifest that only a small proportion of the oxalic acid can be derived from this source. Again, we stated, when speaking of uric acid, that recent views point to the conclusion that even in disease the amount of uric acid formed in the body is not great, so that no one who has observed the enormous amounts of oxalic acid often passed into urine in a single day, or as it exists in calculi, can believe that such abundance could ever come from so small an origin. The view now held is that oxalic acid is derived from a variety of sources, and is found in urine under a variety of clinical and pathological conditions. Thus it may be derived (1) directly from food by the ingestion of substances containing oxalate of lime, such as certain fruits and vegetables, rhubarb, sorrel, tomatoes, onions, turnips, etc. ; (2) indirectly from food, as by incomplete oxydation of the saccharine amylaceous and oleagineous principles of food, which, before their final conversion into carbonic acid and water, yield several intermediary non-nitrogenous acids, of which the chief are glycollic, lactic, and oxalic acids ; (3) from increased tissue metabolism ; this is probably the most frequent cause for the pathological appearance of oxalate of lime in the urine. The urines in these cases are generally

J

of a deep orange colour, of high average specific gravity, with an excess of urea and phosphoric acid, and are usually turbid with mucus and urates, while the deposits of oxalates are not usually persistent, often disappearing for a few days, to return again in great abundance. The explanation of its appearance being, that the process of oxydation within the body, under circumstances of increased tissue metabolism, is only sufficient to reduce a certain quantity of non-nitro-genous fatty acids formed within the body to their lowest term of carbonic acid, and consequently oxalic acid, which is one of the series, appears in the urine. (4) From the mucus of the urinary passages. A very ingenious hypothesis has been advanced by Meckel to account for this formation of oxalate of lime in mucus, by assuming that the mucous membrane of the urinary passages becomes the seat of a specific catarrh. In this catarrh a tough adhesive mucus is secreted, which has a tendency to undergo acid fermentation, and in which oxalate of lime appears when such fermentation occurs. At first this oxalate of lime mucus is of gela-tinous consistence, but gradually it takes up more and more oxalate of lime from the decomposed urine, and thus, growing more and more firm, a stony concretion is at length formed. The large and numerous crystals of oxalate of lime so frequently observed in the urine of persons suffering from spermatorrhœa are most probably derived from the mucus of the genito-urinary passages ; (5) from excess of acid in the system from the increased formation of lactic and butyric acids in intestines, the result of fermentative changes. These acids absorbed from the intestinal canal into the circu-lation being in excess, their reduction into carbonic acid is incompletely performed, and so the intermediate acid, oxalic, appears in the urine in combination with lime. The urine is usually of a pale greenish colour, and the quantity passed in the twenty-four hours normal in quantity and specific gravity. Its chief characteristic is the deposit of crystals of oxalate of lime, which are found most abundantly in the morning urine passed on first rising. Owing to the presence of these crystals causing irritation of the mucous membrane of the bladder, micturition is frequent and urgent, though the quantity of urine passed is not large. Traces of sugar are not infrequently present, and the urine occasionally contains an excess of phosphate of lime. This condition of urine is generally associated with excessive flatulence (flatulent dyspepsia), chiefly connected with the small intestine, and apparently the result of intestinal catarrh, and there is usually great mental depression.

Crystals of calcium oxalate are to be recognised in urine by their peculiar letter envelope shape *a* (Fig. 8), sometimes as mere diamond points, and often as dumb-bells *c*. In the latter case the oxalate of lime is probably derived from the urinary passages. The crystals dissolve in mineral acids, but not in acetic or oxalic acids, which serves to distinguish them from crystalline deposits of the earthy phosphates. They are insoluble in alcohol and water.

Fig. 8.—Crystals of Calcium Oxalate.

Under the blow-pipe they are reduced to carbonate of lime, and the residue effervesces on the addition of acid. (*See* Urinary calculi.)

113. **Phosphoric acid** H_3PO_4. — The amount of phosphoric acid passing out of the system in the course of the twenty-four hours averages from 2·5 grammes to 3·5 grammes, and is distributed among the four bases, potash, soda, lime, and magnesia, in the proportion of about two-thirds combined with the alkaline oxides and one-third with the oxides of the

earths. The alkaline phosphates are extremely soluble, and therefore are never deposited from the urine. On the other hand, the earthy phosphates are only soluble in acid solutions, so that when the urine becomes neutral or alkaline, they are deposited. Thus it happens that a deposit of the earthy phosphates is by no means an indication that they are in excess, any more than the fact that no deposit is present is an assurance that they are being excreted in normal amount. So long as the urine remains acid, a considerable quantity of phosphoric acid may be passing out of the system without giving evidence of its presence, whilst if the urine from any cause becomes alkaline, a deposit at once occurs, although the phosphoric acid may not be eliminated in excess.

(1) *Excess of phosphoric acid, the salts of which are not necessarily deposited.* This condition has long been recognised by writers. The amount of phosphoric acid is enormously increased, often rising from 3 grammes to as much as 8 or 10 grammes in the twenty-four hours. It is chiefly in combination with lime and magnesia, though the soluble phosphates are also increased, but not to the same extent. The discharge of urine is also greatly increased, and so is the excretion of urea. The urine may remain acid throughout the course of the disease, and consequently may be unattended with deposit; but when it becomes alkaline, which it frequently does, bulky white deposits of amorphous calcium phosphate are thrown down. The alkalinity of the urine in these cases is due to an excess of the alkaline carbonate, and when it is persistent the case is usually extremely obstinate. The pathology of this condition is not understood, but it seems to depend on a peculiar state of the nervous system. The increased elimination may be temporary in character and moderate in amount. Such cases usually occur in

persons who have undergone much recent anxiety or mental strain. In this form it often precedes the onset of phthisis. With regard to this, Marcet's analysis of pulmonary tissue in consumption has an important bearing, since he has shown that a considerable reduction of phosphoric acid and potash takes place, in the soluble tissue and nutritive material of the diseased as compared with the healthy lung tissue. Or the increased elimination of phosphoric acid may be excessive and persistent, the disease running a course like saccharine diabetes, into which, indeed, it often passes, but without the appearance of sugar.

(2) *Deposits of phosphate of lime, not, however, necessarily attended with excessive elimination.* In these cases the urine is alkaline from fixed alkali. The urine is turbid or whey-like from the presence of phosphate of lime, which deposits on standing as amorphous granules. Sometimes, however, the phosphate of lime is precipitated in the form of fine acicular crystals (Fig. 9), much resembling some forms of uric acid; they can be distinguished from these by their insolubility in liquor potassæ, and solubility in hydrochloric acid. This form of urine is not infrequently met with in persons convalescent from febrile diseases; and, in certain forms of dyspepsia, deposits of phosphates alternating with deposits of urates and uric acid are often observed. Rickety children often pass urine thus alternating in character. From the frequency with which phosphatic deposit occurs in this disease, it was erroneously held that the phosphates were

Fig. 9.—Crystals of Calcium Phosphate.

in excess; this is now shown to be an error, the amount of earthy phosphates not being increased in rickets. Alkaline urine, when passed into a dirty chamber vessel, often forms an iridescent scum on the surface, owing to the formation of crystals of ammonio-magnesium phosphate, caused by decomposition of urea, which are mixed with granules of calcium phosphate.

(3) *Deposits of ammonio - magnesium phosphate* (*triple phosphate*) $Mg(NH_4)PO_4 + 6H_2O$. We have seen (page 115) that when the urine is alkaline from the presence of volatile alkali (ammonia) we have, in addition to the deposit of calcium phosphate, a crystalline precipitate of magnesium phosphate in combination with ammonia. This salt is usually called triple-phosphate. The crystals are met in different forms, but the most characteristic is that of triangular prisms (*a*, Fig. 10). It is sometimes deposited as feathery crystals (*b*, Fig. 10). This is especially the case when the urine has been artificially rendered alkaline by the addition of ammonia. These crystals are sometimes met with in slightly acid urine. In these cases it is probable that crystals have been formed originally in alkaline urine in the bladder, but which has been rendered slightly acid by subsequent additions of acid urine from the kidney, before the mixed urine is passed; or it may be that, subsequent to emission, the urine which is passed acid undergoes ammoniacal decomposition on its upper surface with the formation of these crystals, while the bulk of the urine remains acid. It has also been suggested that the acid reaction in these cases depends upon some

Fig. 10.—Crystals of Ammonio-magnesium.

salt which reddens litmus paper, but which is not a free acid.

The quantitative estimation of phosphoric acid is performed as follows : Take 100 cubic centimetres of urine and add 10 cubic centimetres of saturated sodium acetate solution ; divide into two portions of 55 centimetres ; each portion, of course, represents 50 cubic centimetres of urine. Reserve one portion in case the process has to be repeated. Heat the other to 100° C., and then from a burette add 3 cc. of standard solution of uranic nitrate; then, after mixing well with a glass rod, touch a drop of solution of potassium ferrocyanide, which is placed in a white dish, with the wet end of the glass rod. If a reddish-brown colour is developed, then too much standard solution has been used, and the process must be repeated with the reserved sample; but this is not likely to be the case, unless the quantity of phosphates is very much below normal. If no brown stain is developed then add 1 cc. more of the standard solution, and after each addition touch the ferrocyanide of potassium solution with the stirring rod ; when the brown stain is developed record the number of centimetres of the standard solution that have been used. Suppose the coloration is given with 18 cc., but not with 17 cc., then take the reserve sample, heat it and run into it 17 cc. of standard solution; then only add $\frac{1}{4}$ centimetre at a time, and you will find exactly the amount required to give the brown reaction, which may be just over 17 cc., or just below 18 cc. Suppose it is 17·5 centimetres, then, as each cubic centimetre of the standard solution is equivalent to ·005 gramme of phosphoric acid, then 17·5 × ·005 gives the amount of phosphoric acid in 50 cc. of urine, from which it is easy to deduce the amount in the twenty-four hours' urine. This process gives the total amount of phosphoric acid in combination with alkaline as well as

the earthy bases. In order to find how much phosphoric acid is in combination with each, we must first
get the total amount of phosphoric acid by the process
described above, and then proceed to determine the
phosphoric acid in combination with the earths separately. This done, we deduct the amount of earthy
phosphates from the total phosphoric acid, the difference being the alkaline phosphates. To calculate the
earthy phosphates separately, take 100 cubic centimetres of urine, and add 20 cubic centimetres of
liq. ammonia, set aside for twenty-four hours.
Filter off the precipitated phosphate of lime and
ammonio - magnesium phosphate, and wash them
thoroughly with dilute liq. ammonia. Then dissolve
the precipitate by means of 5 cc. of strong acetic
acid, and place the filter and the acid solution together
in a beaker and add distilled water up to 90 cc.
Then add 10 cubic centimetres of saturated solution of
sodium acetate. Filter; divide the filtrate into two
equal parts, each of which represents the amount of
earthy phosphates in 50 cc. of urine. Reserve one
portion; with the other proceed, after heating to
100° C., to apply the standard solution of uramic
nitrate, as in preceding process. The number of cubic
centimetres of the solution used will give the amount
of phosphoric acid in combination with earthy bases,
lime, and magnesia, in 50 cc. of urine.

114. **Hydrochloric acid** HCl appears chiefly in
the urine as sodium chloride. Barral has shown that
the quantity of sodium chloride excreted with the
urine does not quite correspond with the amount
taken with food, about one-fifth being decomposed by
acid potassium phosphate to form potassium chloride
and acid sodium phosphate. In acute febrile diseases
the amount excreted by the urine is rapidly diminished, especially in diseases attended with exudation,
as pneumonia, pleurisy, and rheumatic fever; as the

disease declines the chlorides return to the normal excretion, and during convalescence often exceed it. The average amount excreted in the twenty-four hours by a healthy adult may be reckoned as ranging from 5 to 8 grms., varying, of course, with the quantity of food. Chlorine may be estimated volumetrically by two processes : by silver nitrate, or by mercuric nitrate. The latter is the simplest and most reliable ; 50 cubic centimetres of urine, freed from albumin if present, are to be precipitated by the addition of an equal quantity of saturated baryta solution (1 volume barium nitrate, 2 volumes barium hydrate), and then filtered. To the solution add gradually a standardised solution of mercuric nitrate; at first, the white precipitate formed after each addition of the mercuric nitrate solution disappears on shaking. When, however, all the chloride present in the urine has been converted into mercuric chloride, then the mercuric nitrate combines with the urea and forms an insoluble compound. The solution of mercuric nitrate used for estimating chlorine is weaker than that used for estimating urea. It is standardised so that 1 cc. = ·01 grm. sodium chloride. The number of cubic centimetres used to produce a permanent precipitate in 50 cubic centimetres of urine indicates the amount of chloride of sodium present in that amount.

115. **Sulphuric acid** H_2SO_4. — Only a small portion of the sulphur introduced into the body with the food appears in the urine, a considerable portion passing off by the bowels and some by the skin, in the perspiration, hair, nails, and cuticle. Probably, during fasting, the sulphuric acid which appears in the urine is derived from the sulphur of the albuminous constituents of the tissues, and thus the amount of sulphuric acid in the urine may be taken as the measure of the metabolism of the sulphur compounds during fasting. In rheumatic fever and pneumonia the sulphuric acid

is often found considerably increased without any
increased ingestion of sulphur-yielding food. When
carbolic acid is taken or absorbed in large quantities
into the system the sulphates disappear, being converted into sulpho-carbolates. The quantity of sulphuric acid in the twenty-four hours' urine is about
2·5 to 3 grammes. To estimate it quantitatively take
50 cubic centimetres of urine, add a few drops of hydrochloric acid in order to insure complete precipitation
of the sulphate, then add a standardised solution of
barium chloride, 1 cubic centimetre at a time, till a
precipitate of barium sulphate is no longer formed,
then, as 1 cc. of barium solution is equivalent to ·01
gramme of sulphuric acid, the number of centimetres
employed indicates the amount of sulphuric acid in
50 cubic centimetres of urine. The sulphur in the
body, however, is not all oxydised into sulphuric acid,
and a small quantity passes into the urine in a
partially oxydised state. In health about 0·4 gramme
of unoxydised sulphur passes into the urine. In
disease, especially of the liver, this amount is increased. In order to determine the amount, first
ascertain the quantity of sulphuric acid present,
then evaporate an equal portion of the urine and
deflagrate with potassium nitrate; this oxydises the
unoxydised sulphur into sulphuric acid; to ascertain
the amount of this, test with the standardised barium
chloride solution. The amount given will be higher
than when the estimation was made for sulphuric
acid alone. The difference represents the amount of
unoxydised sulphur converted into sulphuric acid.

116. **Hippuric acid** $C_9H_9NO_3$. — This substance is a normal constituent of human urine, the
quantity passed in the twenty-four hours under ordinary circumstances varying from 0·8 to 1 gramme.
Weisman gives 1·17 grms. as the normal daily quantity excreted. The excretion is greatly augmented

by a vegetable diet, and especially by such vegetable substances as benzoic acid, cranberries, blackberries, and plums. Consequently, we are not surprised to find a considerable quantity in the urine of all herbivorous animals; thus cow's urine contains 1 per cent., and horse's urine 0·38. In these animals hippuric acid often undergoes oxydation in the system, and is converted into benzoic acid, which appears in the urine; thus horses at rest pass urine free from benzoic acid and containing the standard quantity of hippuric acid, but when put to hard work the hippuric acid diminishes and benzoic acid appears. Kühne has observed that benzoic acid given to patients suffering from disease of the liver passes unchanged into the urine instead of being converted into hippuric acid, which would have been the case under ordinary circumstances. From this fact he has assumed that hippuric acid is derived from the vegetable aromatic constitu. nts of our food, and the place of their transformation is the liver. Benzoic acid, given internally, is said to diminish the excretion of uric acid, whilst hippuric acid is increased. The excretion of hippuric acid is increased in all febrile affections, also in diabetes. The crystals are semi-transparent rhombic plates, insoluble in cold water, but extremely soluble in solutions of sodium phosphate. Boiled with strong hydrochloric acid, they decompose into benzoic acid and glycocin; the latter crystallises out on cooling. To obtain hippuric acid from urine, evaporate 1,000 cubic centimetres of urine to near dryness, triturate the residue with clean sand, and add 60 cubic centimetres of hydrochloric acid; finally extract with alcohol. The acid alcoholic solution is neutralised with soda ley, and evaporated to a syrupy consistence with a small quantity of oxalic acid, the residue dried in a water bath and treated with a large quantity of ether containing 20 per cent. of alcohol. When the residue is thoroughly exhausted, the alcoholic etherial solution is evaporated and the crystalline residue treated with a solution of milk of lime, and the resulting precipitate removed by filtration. The filtrate is concentrated and hydrochloric acid added; after standing some hours hippuric acid will crystallise out. The crystals are collected on a weighed filter, dried and weighed; the weight gives the quantity of hippuric acid in the amount of urine examined.

117. **Kreatinin** $C_4H_7N_3O$.—This base is constantly present in human urine; according to Neubauer, the quantity passed into the urine in twenty-four hours averages 0·6 to 1·3 grammes. It is derived from the decomposition of kreatin in the blood; in no case has it been obtained as a primary product of decomposition from any of the tissues. Nawrocki has shown, by experiments, that it does not occur in muscular tissue either at rest or when tetanised. The crystals form oblique rhombic prisms, soluble in boiling water and in 12 parts of cold water. It is an extremely powerful base, gives an alkaline reaction with test paper, and forms well-defined basic double salts with zinc chloride and silver nitrate.

ABNORMAL CONSTITUENTS OF URINE.

118. **Albumin.**—Various proteid substances appear in urine, the result of manifold pathological conditions.

(1) *Serum albumin* is the form met with in Bright's disease, in many depraved conditions of blood, in purulent discharges from the genito-urinary tract, and occasionally in persons apparently healthy, under the influence of undue physiological stimuli, such as excessive muscular exertion, excitement, or the ingestion of unsuitable food, etc. It is now generally accepted that the Malpighian bodies are the part of

the kidney by which albumin passes into the urine.
The causes that lead to a transudation of albumin are
generally referred to four conditions : viz., increased
blood pressure, peculiarity of vascular walls, alterations of the renal epithelium, abnormal conditions of
the blood, though no one of these factors alone seems
capable in itself of fully accounting for the phenomenon.   In Bright's disease, albumin is most abundant
in cases originating in acute nephritis.   During the
early stages it is found associated with more or less
blood ; as this clears up, it still continues in large
quantity till the attack has completely passed off,
indeed, often persists after all other evidence of the
kidney affection has disappeared, and the patient has
apparently regained his usual health.   When the
acute attack does not subside, but merges into the
chronic form of nephritis (the large white kidney), the
albumin passed is still very considerable.   In the cirr-
hotic form (small granular kidney), the amount of albumin present in the urine is generally small, and
may be often absent for days.   Professor Grainger
Stewart thinks it probable, indeed, that cirrhosis may
be present as an anatomical change, without the occurrence of any albuminuria.   Dr. Mahomed holds that
there is pre-albuminuric stage in this form of Bright's
disease, in which the urine for a time is free from
albumin.   I have seen several cases tending to support
this view.   In pure lardaceous disease (waxy kidney) the
quantity passed in the early stages is at first small,
but increases as the disease progresses, especially if
any inflammation of the tubules intervenes.   Thus
Professor G. Stewart has watched it gradually increase
from a mere trace till the daily excretion reached one-
twelfth of an ounce.   The albuminuria associated with
depraved condition of blood is generally associated
with some degree of nephritis ; but even when there is
no positive evidence of this, we can readily imagine

how a toxic agency may affect the circulation in the
Malpighian tufts by diminishing the supply of oxygen,
or lessening the amount of nutritive material, and so
allow albumin to escape.   The albumin derived from
pus cannot be distinguished from blood serum.
When there is no kidney disease the case can be distinguished by the absence of tube casts ; when, however, inflammatory disease of the urinary passages
co-exists with Bright's disease, it is impossible to
estimate the amount of albumin derived from each
cause, and our opinion with regard to the condition of
the kidney must be based rather on a consideration
of its functional power as evidenced by the excretion
of urea, and the specific gravity of the urine after it
has been freed from albumin, than from the amount
of this substance present in the urine.   In cases of
temporary, or intermittent, albuminuria, the urine
passed at different periods of the day should be carefully examined, and the examination continued for
some time, till the conditions which lead to its
production are fully understood.   Even after all trace
of albumin has long disappeared from the urine, the
patient should report himself from time to time, for
though in the majority of cases no ill results follow,
still this condition is sometimes a prelude to the
permanent and organic form of albuminuria.

The best plan of procedure in testing for serum
albumin is as follows :

(A) First determine the presence of a proteid or
albuminous substance in urine.   For this purpose any
of the following reagents may be used :* (1) Potassium
ferrocyanide, with citric acid; (2) Potassio-mercuric
iodide, with citric acid ; (3) Mercuric chloride, with

* If the urine is turbid from mucus it must be filtered.   If
the turbidity is due to urates, gently warming the urine will
clear it ; if due to phosphates, the addition of a drop or two of
acetic acid will re-dissolve them.

citric acid; (4) a saturated solution of Picric acid, added
in equal bulk to the urine to be tested ; (5) Concentrated
Nitric acid.   The first three tests can be conveniently
applied by means of Dr. Oliver's test papers, which
are exceedingly handy for clinically testing urine at
the bedside.   In using these tests a citric acid paper
is first dropped into the test tube containing the urine,
and then the special reagent ; if albumin is present, a
delicate haze will diffuse through the fluid.   The
picric acid can be carried in powder in a small box,
and dissolved in water when required.   It is an exceedingly delicate test for albumin, very minute traces
being indicated by it.   It is also useful as a general
urinary test, since it can be made available for testing
for sugar (Sugar, § 119), and for the detection of
peptones in urine.   If uric acid is in excess it is
thrown down by picric acid, and this precipitate may
be mistaken for albumin ; it can be distinguished,
however, by the fact that it is re-dissolved on heating,
whereas the precipitate given with albumin becomes
denser on the application of heat.   Picric acid, too,
has the advantage of precipitating serum albumin
when modified by acid, and also alkali albumin ; at all
events, Dr. George Johnson, who has employed the test
for many years, says he has never met with a case of
highly acid or alkaline urine in which a precipitate
did not occur if albumin was present.   With nitric
acid a zone of coagulated albumin is formed when a
urine containing albumin is floated on the surface,
which is done by placing a few drops of strong nitric
acid at the bottom of a test-tube, and running a little
of the urine carefully down the side of the glass.
Nitric acid as a test has been almost universally
adopted.   It has the disadvantage, however, of being
an awkward reagent to carry about for bedside
purposes, and is not quite so delicate as picric acid.
Like that body, it precipitates uric acid when in excess

from the urine; this precipitation, however, as in the other case, is distinguished by its being readily dissolved by heat.

(B) Secondly, to determine that the proteid body is serum albumin. The above reactions only show that an albumin of some kind or other is present in the urine, and not the variety or the form. This is done by the application of special tests. Now heat is the great distinguishing test for serum albumin, since it is the only albumin (except sero-globulin, *see* page 147) that coagulates at temperature 73° to 75° C. If, then, after having determined the presence of an albumin in the urine, we wish to decide that the body is serum albumin, we must heat the urine. This is best done by nearly filling a test-tube with filtered urine, and applying heat to near the boiling point, when, if albumin is present, a deposit varying from a faint haze to a dense cloud is formed in the hot portion of the tube. The advantage of applying heat this way is obvious, for if there is only the merest haze the difference between the clear cold and the slight turbidity in the heated portion is very readily distinguished, especially if the tube be held to the light, and the dark coat-sleeve placed behind the tube; whereas, if the whole of the urine is heated a slight change may escape observation. In applying the heat test it must be remembered (*a*) that in alkaline or slightly acid urine a cloud of phosphates may be precipitated on boiling them; they clear up, however, on the addition of a drop of dilute acid, or a citric acid paper, whilst albumin does not. If albumin is present as well, both it and the phosphates are thrown down by heat, whilst the precipitate is only partially re-dissolved on adding acid. If any difficulty should occur as to the amount of the partial solution, a fresh sample of urine must be heated very gently and kept for some time just below boiling point, .

75° to 80° C., when the albumin alone coagulates ; on boiling (100° C.) there is an increase in the turbidity, the phosphate being only precipitated at the boiling point. (*b*) Should the urine be alkaline, then the serum albumin may be modified and appear as alkali albumin or casein, in which case no coagulation, or at the most only a slight turbidity, will be given by heat, although the albumin may be present in considerable amount ; if, however, we neutralise the heated layer with a drop or two of dilute acid, coagulation will at once occur ; a citric acid test-paper dipped into the heated layer has the same effect. (*c*) Similarly, if the urine is highly acid, heat will not coagulate the albumin, because it is converted into acid albumin or syntonin ; on neutralising the urine with a drop or two of liquor potassæ precipitation at once occurs ; should, however, the alkali be added in excess, the precipitate is at once re-dissolved. The determination of the amount of albumin present in urine is often roughly performed by judging by the eye the amount of coagulated material deposited in the test-tube in relation to the fluid ; thus it is expressed at one-sixth or one-eighth, etc. This, however, is likely to mislead, unless the specific gravity of the urine is likewise recorded, for a patient may be passing a considerable quantity of albumin in a very dilute but abundant urine, which would of course yield only a small volume of coagula, whilst a scanty but concentrated urine would yield relatively more, though the quantity of albumin present might be absolutely less.

To make an accurate estimation the albumin must be separated and weighed. The procedure is as follows : Take 100 cubic centimetres of urine, place it in a glass beaker, and add two or three drops of strong acetic acid, to render it slightly acid. Place the beaker in a water-bath, 100° C., for about half an hour, frequently stirring to prevent clotting, then set aside to subside.

K

When the coagula have fallen to the bottom of the
vessel decant supernatant fluid into another vessel,
and place the coagulated material on a filter previously
dried and weighed, carefully removing any portion that
may adhere to the glass with a feather to the filter.
Set aside to drain, add from time to time any portion
of coagula that may be deposited from the supernatant fluid that was decanted.   When every visible
fragment of coagula has been transferred to the filter,
place it in the hot-air bath and cautiously dry;
beware of applying heat too urgently at first, or it
will dry lumpy, and consequently take longer to get
rid of all the moisture, since the outer surface will
cake hard and so prevent the moisture from the
interior evaporating.   When it has been in the air-
bath some hours withdraw, cool, and weigh,* and
repeat this process till it ceases to lose weight.
When it does, deduct the original weight of the filter
from the amount, and the difference will give the
weight of albumin in 100 cubic centimetres of urine.
(N.B.—This process answers very well in ordinary
cases, but if the urine is scanty, and the albumin
is abundant, it is necessary to dilute the urine with
twice its volume of water, otherwise the albumin will
separate in clots, and carry down some of the urinary
material, which of course will increase the weight.)

(2) *Paraglobulin and globulin.*—These varieties
of albumin, associated with serum albumin, are met
with in many cases of Bright's disease, chiefly in the
early acute stage, when blood appears in the urine,
and the later stages of chronic white kidney, when
there is much anæmia.   The globulins coagulate by
heat, so that to separate them from serum albumin it

* In weighing precipitates a small beaker half-full of strong
sulphuric acid should always be placed in the case of the
weighing machine, to keep the air of the chamber dry, otherwise
moisture will be absorbed and weight increased.   The filter should
always be allowed to cool in this chamber.

is necessary to employ a reagent that does not affect that body. This is effected by precipitating with magnesium sulphate; the filtrate is heated to 75° C., which throws down the sero-albumin, by passing a stream of carbonic acid through the urine, which should be diluted three or four times its bulk with distilled water. I am not aware that the globulins ever appear in urine unless accompanied with serum albumin; the conditions under which both pass out of the kidney seem to be identical, though in some forms of temporary albuminuria paraglobulin appears to be in excess of the serum albumin.

(3) *Fibrin* is met with in urine, associated with chylous urine, from which it separates as a light gelatinous clot. It is known by its power of decomposing hydrogen peroxide. After hæmaturia moulds of the urinary tubes, consisting of decolorised fibrin and fatty matter, are sometimes passed, they also cause effervescence with hydrogen peroxide.

(4) *Parapeptone*, or pro-peptone, sometimes appears in urine. This body is one of the intermediate products of gastric and pancreatic digestion. Before peptone is arrived at, according to Kühne, anti-albumose and hemi-albumose are formed. Of these, anti-albumose corresponds to acid albumin or syntonin, whilst hemi-albumose is probably equivalent to the so-called c peptone of Meissner, which has been identified with the peculiar form of albumin discovered by Bence Jones in the urine of a case of osteo-malacia. This body gives a precipitate with nitric acid or picric acid in the cold, but which re-dissolves when heated to 70° C.; this form of albumin is frequently associated with the presence of true peptone in the urine.

(5) *Peptones.*—It has long been known that these bodies often make their appearance in urine, but the subject has not received in this country the

attention it deserves. Frerichs, Schultzen, and Riess have met with them in the urine of cases of acute yellow atrophy and phosphorus poisoning ; Eichwald in acute parenchymatous nephritis ; Petri found them in twenty-eight cases out of forty-one cases tested ; Gerhardt found them in the urine of patients suffering from diphtheria, tertiary syphilis, pneumonia, typhus, and typhoid fever; in some of his cases they preceded the coming-on of albumin in the urine. Gerhardt's observations have been confirmed by Obermüller. I have recorded* three cases, in one of which there was slight temporary albuminuria, and in two, though there was no albumin, the patients presented the appearance, and had many of the symptoms, that would lead one to expect granular or contracted kidney. The test for their presence is the peculiar rosy red they give with alkaline solutions of cupric sulphate in the cold. To bring this out clearly, place about a drachm of Fehling's solution in the bottom of a test-tube, and float an equal quantity of urine on the surface ; where the two fluids meet a zone of phosphates will be deposited, above which, if peptones are present, a red halo will develop. If peptones alone are present, then the red is of a rosy or pink tint. If there is also much albumin, then the red is more of a violet hue. Now if, by means of a pipette, a drop or two of picric acid is allowed to fall in, the red coloration turns to deep-red, then to reddish-yellow, and finally becomes yellow. It is a difficult matter to obtain these bodies for examination. The plan that has been usually adopted has been to precipitate them with alcohol. Now albuminous peptones are not quite insoluble in alcohol, whilst mucin is freely precipitated by it; so that, unless mucin be previously removed, this substance may be mistaken for peptone.

* *British Med. Journal*, May 12th, 1883.

Again, if the urine contain albumin, it is very difficult to separate it completely, a small quantity always
remaining even after repeated coagulation by heat.
The process I have devised is a modification of that
adopted by Schultzen, Riess, and Hofmeister. It is
as follows :—500 cubic centimetres of filtered urine are
to be placed in a glass vessel, and 10 cubic centimetres
of strong acetic acid added. If mucin is present in
solution a turbidity ensues. The urine is allowed to
stand twelve hours, by which time the precipitated
mucin falls to the bottom of the vessel. A drop of
acetic acid should then be added to the supernatant
fluid to see if all the mucin has been thrown down ;
if this drop causes a cloudiness when it is allowed
to fall, then more acetic acid (5 cc.) must be
added, and the precipitate allowed to collect ; and
this process must be repeated till acetic acid causes no
cloudiness when added. The clear supernatant fluid
is then to be decanted off and filtered. To this add,
drop by drop, a concentrated solution of ferric
chloride, till the solution has acquired a permanent
red colour. Then carefully neutralise the solution
with a concentrated solution of sodium carbonate.
Allow the precipitate to subside ; then pour off the
supernatant fluid and filter it. The filtrate ought now
to be free from albumin, and give no reaction with
potassium ferrocyanide and acetic acid. The filtrate
is now to be evaporated to half its bulk, and 250
cubic centimetres of absolute alcohol added whilst the
liquid is yet warm. If peptones are present, a
brownish precipitate is the result. The whole should
be kept in the water-bath (100° C.) for twenty-four
hours, alcohol being added from time to time. When
a precipitate is no longer formed, the brown substance
must be removed to another vessel, and boiled with
alcohol for twelve hours. The precipitate is then
collected, dried, washed with ether, dissolved in water,

re-precipitated by alçohol, again collected, dried, and washed by ether. At length, after repeating this process three or four times, a grey-yellowish powder is obtained, which is hygroscopic, is easily soluble in water. The solution is neutral in reaction, gives no precipitate with potassium ferrocyanide and acetic acid, turns the plane of polarised light to the left. The powder, when heated to 180° C., evolves ammonia. The alcoholic extracts, exhausted with ether, yield to the etherial solution a brown, thin residue, out of which thin crystals of tyrosin separate.

119. **Sugar (glucose).**—Healthy human urine contains, as has been proved by Pavy, minute traces of glucose ; but, in certain unnatural states of the system associated with disturbance of the hepatic function, a larger amount passes into the urine, inducing a condition known either as glycosuria, or diabetes, according as it acquires a more or less marked saccharine reaction and is temporary or permanent in its character. The nature of the perversion of liver function which leads to the increased passage of sugar into the blood, and hence into the urine, and the difference of degree observable in the various forms of diabetes and glycosuria, are considered in the chapter on the digestive organs (§ 145).

Various substances produce a reaction with glucose; but the tests used for clinical purposes are, (1) the alkaline copper test ; (2) the yeast, or fermentation, test ; (3) the indigo carmine test ; (4) the picric acid and liquor potassæ test ; (5) polarised light.

(1) *The alkaline copper test.*—Alkaline solutions of glucose possess the power of reducing cupric salts to cuprous. This property is made use of in detecting sugar in urine. The test solution in general use is that of Fehling. It is made by weighing 34·63 grains of pure crystallised cupric sulphate, and adding distilled water up to one litre. One cubic centimetre of

this solution is equivalent to ·005 grm. of sugar. The alkaline solution is prepared by dissolving 173 grms. of pure crystallised sodio-potassium tartrate and 80 grms. of potassium hydrate in distilled water, and filling up to the measure of one litre. The copper and alkaline solutions must be kept in separate bottles. Thus prepared, the test is available for quantitative as well as qualitative purposes. In testing for sugar, we place equal quantities (one centimetre of each solution) in a perfectly clean test-tube, and boil ; then set aside for a few minutes to see if the solution is in good condition and has not been impaired by keeping. If in good order, it remains perfectly clear and retains its deep blue colour. If faulty, it becomes turbid and thick, in which case it is not available. If, however, it is good, run a drop of the suspected urine (which must be freed from albumin, if present) down the side of the tube, and gently heat. If sugar is abundant, a yellow precipitate, turning red, will be formed. If no precipitate is formed, then add a drop or two more urine, and so on till a bulk equal to the amount of test has been added. If then there is no precipitate, we may be sure that sugar is absent. When the sugar is in extremely minute quantity, the reduction may be very slight, and, instead of a red or reddish-yellow, the precipitate is greenish-yellow, caused by the admixture of the yellow urine and a little reduced copper with the blue of the Fehling, and rendered opaque by the presence of precipitated phosphates. On standing, however, a few grains of cuprous oxide will deposit at the bottom of the tube ; this distinguishes it from the greenish coloration caused by inosite. Other substances, as uric acid, kreatinin, etc., found in the urine besides sugar, possess when in excess the property of reducing cupric salts ; we must be on our guard, therefore, against taking these for sugar. In typical cases of diabetes or glycosuria, the

reaction in a few drops of urine is too marked to lead
one into error ; but, when the reduction is slight, the
question arises whether it is caused by uric acid or
sugar. To decide this question, add to the urine a
solution of lead acetate ; filter ; and then test the
filtrate with the copper solution. If it does not now
give the reaction, then the former reduction was due
to uric acid ; if, on the other hand, the cuprous oxide
is still thrown down, we may be sure the reduction is
due to glucose. In employing the test, care must be
taken only to apply gentle heat, since nitrogenous
matter in urine in the presence of alkalies gives off
ammonia, which holds the cuprous oxide in solution
and prevents its deposition. Again, the urine should
only be added drop by drop, since excess of glucose will
hold the cuprous oxide in solution. The quantitative
estimation of sugar by means of Fehling's solution is
performed as follows : Into a porcelain basin, capable
of holding 500 cubic centimetres, place 50 cc. of
distilled water, and to this add, carefully measured by
means of a pipette, 10 cc. of the copper solution
and 10 cc. of the alkaline solution, prepared as
directed above ; then take 5 cc. of the diabetic urine,
freed from albumin if present, and dilute it up to 100
cc. with distilled water, and place 50 cc. of this in a
Mohr's burette, reserving the other 50 co. in case of
accident or of more being required to complete the
process. Now very gradually heat the contents of
the porcelain dish to just the boiling point, and then
run a few drops of the diluted urine from the burette.
This at first is of a muddy colour, but after each
addition becomes redder and redder as more of the
copper salt is reduced. After each addition the
porcelain dish is tilted a little, to see if the edge of
the fluid has become colourless. When this is the
case, a few drops of the contents of the porcelain
basin must be withdrawn by means of a pipette, and

passed through a small filter into a test tube containing a small quantity of potassium ferrocyanide solution and a drop or two of acetic acid. If no brown
coloration is given, the process is complete; but, if
there is, then more of the diluted urine must be carefully added to the contents of the porcelain basin, till
the ferrocyanide solution gives no reaction. The calculation is made as follows: Read off the number of
centimetres of dilute urine used from the burette;
say, for sake of example, they amount to 57 cubic
centimetres. Now, as the urine was diluted to one-
twentieth of its volume, 57 cc. of diluted urine are
equivalent to 2·85 cc. of the diabetic urine. Again, 1
cc. of the copper solution is equivalent to ·005 gramme
of sugar; and, as 10 cc. were employed for reduction, then 2·85 cc. of diabetic urine contained ·45
gramme of sugar ; and, supposing the patient passed
3750 cubic centimetres of urine in the twenty-four
hours, then $\dfrac{3750 \text{ cc.} \times \cdot 05 \text{ grm.}}{2 \cdot 85} = 65 \cdot 78$ grammes
of sugar passed in the twenty-four hours. In other
words, the same result is obtained if we divide the
twenty-four hours' urine by the number of centimetres
of the dilute urine used from the burette; thus
$\dfrac{3750}{57} = 65 \cdot 78$ grammes of sugar.

(2) The *fermentation test.*—Yeast added to a
solution of glucose, and kept at a temperature of 25°
to 30° C., speedily undergoes vinous fermentation ; as
the sugar is converted into carbonic acid, and this
passes off into the atmosphere, the solution loses weight.
This fact has been made use of for the quantitative
estimation of sugar for clinical purposes. Two equal
portions of urine (4 ozs.) are placed in two 6-oz. medicine
bottles, and in one is placed a fragment of baker's
yeast, about the size of a small bean, and the mouth
lightly plugged with cotton wool. The two bottles

are to be kept in a warm place for twenty-four hours. In the bottle in which the yeast is placed fermentative action soon commences and carbonic acid formed, which passes off through the cotton wool. At the end of twenty-four hours the specific gravity of both bottles is taken, and the difference represents the amount of sugar, each degree of specific gravity lost in the bottle which has undergone fermentation representing *one grain of sugar in each ounce* of the twenty-. four hours' urine; then if the patient passed 260 ounces of urine in the day, and the difference in the specific gravity amounts to 7 degrees, then 260 × 7 = 1820 grains of sugar passed by the patient in twenty-four hours. If French measures are employed instead of English, then each degree of specific gravity lost represents 0·2196 gramme of sugar in every 100 cc. of urine.

(3) *Indigo carmine test* is based on the fact that indogotine, the colouring matter of commercial indigo, when heated with an alkali in the presence of glucose and certain carbohydrates, is converted into indigo white, and which is capable of reconversion, under the influence of oxygen, back into indogotine. The change is represented in the following equations: Indogotine $2(C_8H_5NO) + H_2 = C_{16}H_{12}N_2O_2$ indigo white. Dr. Oliver has recently made this test readily available by means of specially prepared test-paper. A strip is placed in a test-tube and covered with distilled water, and heated till a blue solution is formed; a drop of diabetic urine is then introduced and heat applied, care being taken not to shake the solution or allow it to boil. The solution gradually becomes violet, then purplish, then orange-red, then reddish-yellow, and finally straw-coloured. Now on ceasing to heat, and shaking the test-tube, the liquid passes back through the different colours into the original blue. This test is likely to prove a valuable supplement to the other tests for sugar. The copper test, as is well

known, is not reduced by all forms of sugar, nor do
all kinds ferment readily with yeast; now as the indigo
reaction is given by many forms of carbohydrate,
it may be thus made available for distinguishing
between those forms of sugar sometimes present in
urine which give no reaction with copper, and which
do not readily ferment, and so help to distinguish those
cases from true glycosuria.

(4) *Picric acid test.*—When an alkaline solution of glucose is heated with picric acid, the liquid
assumes a deep red-brown colour, due to the formation
of picramic acid. This affords an extremely delicate
test for glucose in urine, and it has also an additional
advantage for clinical purposes, since picric acid is
a delicate test for albumin, and also that the presence
of albumin does not interfere with reaction for sugar.
In applying the test, add an equal bulk of saturated
solution of picric acid to the urine; if albumin is present a cloudy precipitate will form; then add a few
drops of liquor potassæ, and gently apply heat, the
solution will gradually acquire a deep red-brown colour.
Nearly all urines treated in this way become darker in
colour, but the coloration in no way approximates
to that yielded by even the most minute trace of
sugar. Dr. George Johnson, who has paid much
attention to the application of the picric acid test for
clinical purposes, has devised an exceedingly ingenious
method of quantitatively estimating the amount of
sugar by the depths of colour yielded by this reaction
as compared with a standard colour for comparison.
His method is likely to be largely used for clinical
investigations, as it can be quickly performed.*

Take a fluid drachm of a solution of grape-sugar,
in the proportion of a grain to the fluid ounce; mix
it with half a drachm of liquor potassæ (*P.B.*), and ten
minims of a saturated solution of picric acid; and

* *Brit. Med. Journal*, March, 1883.

make up the mixture to four drachms with distilled water. The mixture is conveniently made in a boiling tube, ten inches long and three-fourths of an inch in diameter, which may be marked below at the height of two and four drachms. With a long boiling-tube there is little risk of the liquid boiling over; and the steam, condensing in the upper cool part of the tube, flows back as liquid, so that there is little loss by evaporation. The liquid is now raised to the boiling point, and the boiling is continued for sixty seconds by the watch, so as to insure the complete reaction between the sugar and the picric acid. During the process of boiling, the pale yellow colour of the liquid is changed to a beautiful claret red.

The liquid having been cooled, by cautiously im-mersing the tube in cold water, and it having been ascertained that its level is that of the four-drachm mark on the tube, or, if below the mark, it having been brought up to it by the addition of distilled water, the colour is that which results from decomposition of picric acid, by a grain of sugar to the ounce, four times diluted; in other words, it indicates one-fourth of a grain of sugar to the ounce; and this colour is a con-venient standard for comparison in making a volumetric analysis. The picramic acid solution, however, on exposure to light, even for a few hours, becomes paler; but the colour may be exactly imitated by a solution of ferric acetate, with a slight excess of acetic acid and an excess of ferric chloride. The iron solution we have found to retain its colour unchanged for a fort-night, even when exposed to a strong light; and we expect that, when light is excluded, it may be kept for an indefinite period; and it is, therefore, a conve-nient standard for comparison.

If, now, a drachm of a solution of grape-sugar, containing two grains to the ounce, be mixed with the same quantity of liquor potassæ (half-a-drachm) as before, but with double the amount of picric acid (*i.e.*, twenty minims), and made up to four drachms in the boiling tube, the result of boiling the mixture as before, for sixty seconds, will be the production of a much darker colour than when the one-grain solution was acted upon ; but if now the dark liquid be 'diluted with its own volume of water, the colour will be the same as that of the one-grain solution.

The dilution is accurately done in a stoppered tube, twelve inches long and three-quarters of an inch in diameter, graduated into $\frac{1}{10}$ and $\frac{1}{100}$ equal divisions (Fig. 11). By the side of this tube, and held in position by an S - shaped band of metal, is a stoppered tube of equal diameter, and about six inches long, containing the standard iron solution.

Sufficient of the dark saccharine liquid to be analysed is poured in to occupy exactly ten divisions of the graduated tube. Distilled water is then added cautiously, until the colour approaches that of the standard. The level of the liquid is then read off and noted. A more exact comparison of the saccharine liquid with the standard is made by pouring into a flat-bottomed colourless tube, about six inches long and an inch in diameter, as

Fig. 11.

much of the standard as will form a column of liquid about an inch in height, and an exactly equal column of the saccharine liquid in a precisely similar tube. The operator then looks down through both tubes at once, one being held in each hand, upon the surface of a white porcelain slab, or a piece of white paper. In this way a slight difference of tint is readily recognised, and if the liquid to be analysed be found to be darker than the standard, it is returned to the graduated tube, and diluted until the two liquids are found to be identical in

colour, when the final reading is taken. The saccharine
liquid having been diluted four times before it was
boiled, a colour equal to that of the quarter-grain
standard would indicate one grain of sugar per fluid
ounce. If further dilution were required (say from
ten to twenty divisions) the proportion of sugar would

Fig. 12.—Polarimeter.

be two grains per ounce, and so on to thirty or forty
or upwards, or to intermediate divisions. Thus dilution from ten to thirty-five divisions would indicate
3·5 grains of sugar per ounce.

(5) **Polarimetry.**—Glucose possesses the property of rotating polarised light toward the right.
This property has been made use of to determine the
amount of sugar present in solution, by the amount of
deviation observed. This is done by an instrument
termed a saccharometer, represented at Fig. 12. Light
is admitted through a Nicol's prism or *polariser* at *b*,
and falls on another prism, the *analyser*, at *a*. Now
if these two prisms are arranged so that no light passes

through the analyser or second prism, which is done by turning it round to a certain degree by means of screw *d*. The instrument being so adjusted that no light passes through the second prism, the tube *c c* is then filled with a solution of glucose, and placed between the two prisms *a* and *b*. Immediately light passes again through the second prism *a*, and this must be turned by means of *d* through a certain angle till the light can again be stopped. The magnitude of this angle is read off on scale *e*. Now the magnitude is in direct proportion to the length of the tube and the quantity of sugar in solution. If, therefore, we know the specific rotatory power of the substance submitted for analysis $[a]_D$, which in the case of glucose is $+ 57 \cdot 6$, and have ascertained the length of the tube *l*, and the magnitude of the angle

of deviation, then $\dfrac{a}{[a]_D \times l} = x$, the weight in grammes

of the substance present in 1 cc. of the solution employed. In examining diabetic urine (freed from albumin, if present), 10 cc. of solution of lead acetate are added to 100 cc. of the urine, in order to remove the colouring matter of the urine, and the solution filtered; the liquid is then placed in tube, the length *l* of which, in decimeters, has been ascertained, and the

angle of deviation *a* read off, then $\dfrac{a}{57 \cdot 6 \times l} = x$ grms.

$$[a]_D$$

of glucose in each cc. of urine.

120. **Blood** appears in the urine under a variety of conditions. (*a*) It may come from any part of the genito-urinary tract, the result of local disease or injury, as in acute nephritis, calculous disease, parasites, cancer, tubercule, etc. ; (*b*) or it may result from certain depraved conditions of the blood itself, as in scurvy, purpura, and the hæmaturia that often attends eruptive and continued fevers of malignant type ; (*c*)

or from simple passive congestion, such as occurs in the
obstructive forms of heart disease; (d) from tempo
rary disturbance of the renal circulation, as in inter
mittent fever, or from mental emotion, and the like.
The quantity of blood passed into the urine is very
variable. It may be so small as only to give a smoki-
ness to the urine, or so great as to colour the urine
deep red, and to separate into large coagula in the
urinary passages. It must be remembered, however,
that a very little blood is capable of giving a very deep
coloration to urine. In some experiments I made in
1873, at the laboratory of Charing Cross Hospital, and
published in the *Lancet*, I found that only one part of
blood gave a decided smoky tint to 1500 parts of normal
urine, whilst 1 part in 500 gave a bright cherry colour.
Considerable hæmorrhages, therefore, are best judged by
the amount of coagula rather than by mere intensity of
colour. When blood appears in urine and the blood
corpuscles are recognised under the microscope, the
condition is termed hæmaturia; when no corpuscles
are to be found, but only the colouring matter, then
it is spoken of as hæmatinuria.

(1) *Hæmaturia.*—The character of the hæmorrhage, together with the general and special symptoms, is
usually sufficient to indicate the part of the genito-
urinary tract from whence it is derived, thus :—

(a) Acute Nephritis.—Smoky to dark brown urine
persistent for some days, with granular and blood
casts, and excess of albumin.

(b) Renal calculus.—Often deep red from excess
of blood, increased by movement, and passing off
rapidly if the patient is kept quiet in bed, so that
only a few blood corpuscles can be seen in the urine.
Generally accompanied or immediately following a
severe attack of colic; retraction of testicle on side
affected. Vesical calculus; hæmorrhage generally
follows undue movement, especially jolting; bladder

symptoms, prominent; detection of stone in bladder by sound.

(c) Cancer of kidney.—Hæmaturia very abundant, with large coagula, and repeated at irregular intervals, generally tumour in loin. Cancer of bladder, frequent and profuse hæmorrhage, cancer cells in urine, pain referable to bladder, and a tumour may be discovered with sound.

(d) Morbid conditions of the blood.—Hæmorrhage often profuse, but rarely attended with formation of clot. General constitutional symptoms manifest.

(e) Intermittent hæmaturia.—The blood passes at very irregular intervals, is generally associated with a considerable quantity of albumin and a definite rise of temperature. In these cases there is usually a history of ague, if the disease is not actually in progress; it is sometimes associated with intermittent chyluria or gout.

(2) *Hæmatinuria.* — In these cases only the colouring matter of the blood is present, no blood corpuscles, or only a few, are to be found. The attacks come on in paroxysms, attended with a chill, and generally accompanied with some degree of nausea and slight jaundice. The urine has a port-wine colour, and is usually passed clear. On standing it deposits a granular sediment, consisting of a few tube casts and fibrinous cylinders, epithelium, crystals of calcium oxalate, in some cases crystals of hæmatin have been observed. From the fact that the spectrum of this kind of urine invariably shows the characteristic bands of hæmoglobulin, many writers designate the disease hæmo-globinuria. But in addition, however, to the bands in D and E characteristic of hæmoglobulin, there is invariably a third present near C, which corresponds with that given by methæmoglobin in acid solutions. (*See* Fig. 4, § 95.) Formerly methæmoglobin was supposed to be per-oxy-hæmoglobin,

L

in which case by reduction methæmoglobin ought to yield oxy-hæmoglobin, but it yields, however, reduced hæmoglobin. This and other considerations make it probable that methæmoglobin, instead of containing more oxygen, actually contains less than oxy-hæmoglobin, and that as regards iron, whilst this body is at its minimum in oxy-hæmoglobin, it is at its maximum in methæmoglobin. These considerations, therefore, make it probable that the oxy-hæmoglobin is undergoing a downward change into hæmatin. The pathology of the disease is still obscure, though the interesting researches of Jaffé and MacMunn on the origin of the urinary pigments seem to throw some light on it. If the effete hæmoglobin is converted in the liver into hæmatin and then into bile pigments, one of which, urobilin, is oxydised into choletelin, which appears in the urine ; and if this conversion does not always take place, it may be owing to some functional disturbance of the liver, the effete hæmoglobin may escape conversion in the liver, and be eliminated by the kidney, as hæmoglobin undergoing the downward change into hæmatin before conversion into choletelin (§ 100). That there is some disturbance of hepatic function is indicated by the jaundice attendant on this condition.

The tests for blood in urine. When due to hæmaturia, the detection of blood corpuscles by the microscope affords the best and most positive indication of the presence of blood. These, however, vary in shape. If the urine is moderately acid they retain their natural form a considerable time, but they become jagged at their edges, lose colour, and no longer adhere together. If the blood corpuscles become dissolved, then we may try Heller's hæmatin test, which consists in boiling the urine with a concentrated solution of caustic potash, when the

phosphates carry down the hæmatin as a reddish-
brown precipitate, which by transmitted light has a
greenish tint ; and which precipitate, treated with
sodium chloride and acetic acid, yields crystals
of hæmin (Fig. 3, § 95). The guiacum test is
best applied by filling a test-tube one-third with
tincture of guiacum and adding a few drops of the
suspected urine, then float about a drachm of etherial
solution of peroxide of hydrogen on the surface, and
at the line of junction, if blood is present, a delicate
blue colour is developed. This test, however, is not
to be depended on, except as a general one, since often
extraneous substances besides blood may give the
reaction in urine. If any doubt exists, the spectroscopic examination will reveal characteristic bands of
oxy-hæmoglobin, reduced hæmoglobin, methæmoglobin,
or hæmatin, according to circumstances, and the
degree of decomposition that has occurred.

121. **Bile.** — The conditions which lead to the
appearance of bile in the urine are discussed in
chapter v., § 138. Urine containing bile varies from a
deep brownish-red to the colour of London porter.
A linen rag dipped in it acquires a yellow stain.
This coloration is due to bile pigment. The reaction
for this is readily shown by placing on a white
plate or dish a few drops of the urine, and
near it a few drops of nitric acid to which
one drop of sulphuric acid has been added (to form
free nitrous acid), and then allow the two fluids to
mix. If bile pigment is present, a play of colour is
observable, of which the green tint is characteristic of
bile pigment (*Gmelin's test*). Bile acids are present,
as well, in the urine of some cases of jaundice.
They are detected by adding a few grains of glucose
to the urine, and then allowing the solution to run
down the side of a test in which a little concentrated sulphuric acid has been placed ; a purple

reaction (*Pettenkoffer's test*) will develop if they are present. As some other substances give this reaction, if there is any doubt the bile acids must be separated from the urine by evaporating it to a thick syrup, extracting with ordinary alcohol, evaporating the alcoholic solution, and treating the residue with absolute alcohol. Evaporate this solution, and dissolve the residue in distilled water, and precipitate the solution with neutral and basic lead acetate. The precipitate is dissolved in water, and decomposed with hydrogen sulphide, and filtered. The filtrate is treated with excess of sodium carbonate, and concentrated. On standing, long needle-like crystals of glycocholate of soda will form, with oily globules of taurocholate of soda.

122. **Chylous urine.**—In the disease known as chyluria the urine has a milky appearance, sometimes slightly tinged with blood, and yields a delicate fibrinous clot. On standing this clot comes to the surface of the urine, and forms a distinct, jelly-like layer. This clot possesses the property of fibrin, of decomposing peroxide of hydrogen. Besides the addition of this abnormal matter, the urine is little altered in its other characters. After the separation of the clot, the blood, the extraneous albumin, and the removal of the fatty matter by extraction with ether, I have found the urea normal in proportion, though the amount of water is relatively increased. The chylous matter is apparently derived from the lymphatics, probably due to some lesion of those connected with the kidney. The disease is sometimes persistent, but frequently it is occasional and intermittent. There are cases on record in which the urine was observed milky on a few occasions only, but never afterwards made its appearance. The most completely recorded case of typical persistent chyluria, associated with filaria in the blood, is by

Dr. Stephen Mackenzie.* Brieger† has given some very complete analyses of the night urine; two, the maximum and minimum, are here appended :

	Maximum in 100 parts.	Minimum in 100 parts.
Fats	0·725	0·06
Albumins	0.798	0·581
Urea	3·4	3·7
Uric acid	0·03	0·03
Sodium chloride . . .	1·7	1·4
Sulphates	0·22	0·23
Quantity of urine . . .	400cc.	300cc.
Specific gravity . . .	1·016cc.	1·025cc.

Both urines contained peptones and traces of indican. The fatty matters consisted of ordinary fatty matter, lecithin, and cholesterin. To examine a chylous urine, filter off the coagulum, and exhaust it thoroughly with ether, water, and alcohol till quite free from fat and albumin; dry and weigh. Agitate the urine with successive quantities of ether till no more fat is taken up. Evaporate the etherial solution in a weighed platinum capsule, weigh the dried residue, which gives amount of fatty matter. (For separate determination of the nature of these fatty matters see § 99.) The urine freed from fatty matters can then be examined quantitatively for albumin (§ 118), peptones (§ 118, No. 5), urea (§ 110), uric acid (§ 111), chlorides (§ 114), sulphates (§ 115), phosphates (§ 113).

Apart from chyluria the urine often assumes a milky appearance, from the presence of fatty matters

* Path. Soc. Trans., vol. xxxiii., p. 394.
† Zeitsch. Phys. Chemie, p. 411, 1880.

in a state of extremely fine subdivision; the urine in these cases is generally slightly albuminous, and often gives the peptone reaction with cupric sulphate. In Bright's disease, fatty matters may appear in the urine, apparently derived from the degenerated epithelium of the tubules. I have met with a small quantity of fatty matter in the urine of a patient dying of acute diabetic coma. Plates of cholesterin are sometimes met with as a urinary deposit.

123. **Cystin** $C_3H_7NSO_2$. — According to Dr. Bence Jones, cystin is constantly being separated in the healthy organism, immediately undergoing transformation into sulphuric acid, carbonic acid, and urea. Whenever this chemical transformation is arrested cystin appears in the urine. As the composition of cystin is $C_3H_7NSO_2$, the proportion of nitrogen to carbon is four to twelve; in uric acid $C_5H_4N_4O_3$ it is four to five; and in urea CH_4N_2O four to two : therefore twelve, five, and two are the indices representing the different amounts of suboxydation in cystin, uric acid, and urea respectively. Dr. Bence Jones thus regards cystin as representing the smallest degree of the oxydation of the albuminous principles, in the same way that sugar in diabetes represents the least degree of oxydation of the amylaceous principles. Cystin, however, as a urinary deposit is extremely rare.

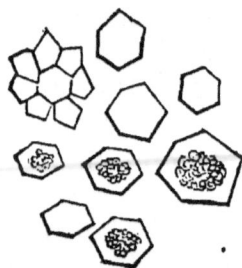

Fig. 13.—Cystin Crystals.

When it appears it forms a whitish or fawn-coloured sediment, which is dissolved by ammonia, and from which on evaporation it is deposited in hexagonal plates (Fig. 13), and which are insoluble in acetic acid. Heated in a solution of lead acetate, a precipitate of lead sulphide is formed, owing to the sulphur contained in cystin. The pathological

significance of cystin in urine is not yet determined.
I have met with it in the urines of strumous children, and in adults suffering from disease of the liver.
It sometimes is retained in the urinary passages and
forms a calculus.

124. **Xanthin** $C_5H_4N_4O_2$ is a constituent of certain
rare urinary calculi, and Dr. Bence Jones has recorded an interesting case of xanthin gravel in a boy
aged nine years. The xanthin calculus removed by
Langenbeck was also from a boy. Dr. Bence Jones
considers that the xanthin diathesis will be generally
found to occur in youth, as it is in the early period of
life the greatest chemical variations of the body are
to be expected, and the most imperfect oxydation of
xanthin into uric acid most likely to occur. To
separate it from urine, add baryta water till a precipitate is no longer thrown down; filter, and evaporate
the filtrate to a syrup, and allow it to crystallise.
The mother liquor, after the removal of the crystals,
is boiled with cupric acetate, and the precipitate thus
formed is removed by filtration,
washed, and dissolved in warm nitric
acid. This acid solution is precipitated by silver nitrate, and the resulting precipitate washed and crystallised from dilute nitric acid, and the
crystals washed with ammoniacal
silver solution, and suspended in
water. The aqueous solution is to
be decomposed with sulphydric acid,
filtered, and the filtrate evaporated; the residue yields xanthin.
Xanthin forms white scales, somewhat resembling beeswax in appearance. Deposited spontaneously from
urine it occurs in lemon-shaped plates (Fig. 14, *a*);
these, dissolved in hydrochloric acid and the solution

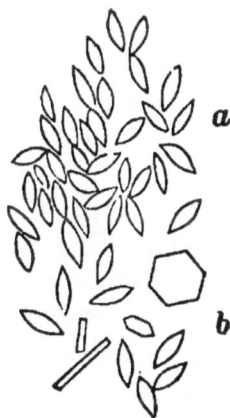

Fig. 14.—Xanthin
Crystals.

evaporated, yield prismatic and hexagonal crystals (b). Xanthin is insoluble in water, alcohol, and ether; soluble in alkaline solutions, from which it is deposited by a current of carbonic acid gas; and in strong mineral acids. Burnt in air, it gives off the odour of scorched hair. Evaporated with nitric acid on platinum foil, and the residue moistened with liquor potassæ, it yields a dark purple colour. Xanthin gives white precipitates, with mercuric chloride and silver nitrate. Dissolved in hydrochloric acid, it gives with platinic chloride a yellow crystalline precipitate.

125. **Hypoxanthin** $C_5H_4N_4O$ has been found in the urine of patients suffering from leucæmia; to separate it, add baryta water and filter off the precipitate. To the filtrate ammoniacal solution of silver nitrate is added, and the greyish-white precipitate collected on a filter and washed. The precipitate is then to be suspended in water, decomposed with sulphydric acid, the mixture boiled for some time, and filtered while hot. The filtrate is next evaporated to dryness, the residue contains uric acid, xanthin, and hypoxanthin. To separate the uric acid and xanthin, the residue is to be dissolved in dilute sulphuric acid, boiled, and filtered whilst hot; to the filtrate, when cold, mercuric nitrate is added and filtered. To the filtrate some ammoniacal solution of silver nitrate is added; the precipitate consists of hypoxanthin nitrate and silver oxide. This is decomposed with sulphydric acid and hypoxanthin is precipitated as a white, imperfectly crystalline powder, rather more soluble in water and alcohol than xanthin. It dissolves freely in acids.

126. **Leucin** $C_6H_{13}NO_2$ has generally been met with in the urine in cases of acute yellow atrophy of the liver, in cirrhosis of that organ, and in severe cases of small-pox and typhus. Dr. Anderson, however, has found leucin somewhat frequently in urine

under less severe conditions. He believes that both
leucin and tyrosin are found in the urine under numerous different pathological conditions, whether affecting
the liver intrinsically, or from without; and that as
often, or as soon as the patient recovers, these substitution products for urea first diminish in amount and
then disappear (*Med.-Chir. Soc. Trans.*, vol. lxiii.,
p. 245). Without entirely agreeing with all the
author's conclusions, I believe we should find leucin
more frequently if we looked for it. Leucin, as well
as tyrosin, has an interest in consequence of its
formation during pancreatic digestion. When present
in urine, it is only required to evaporate about
five ounces of that fluid to a thin syrup, and, when
cold, leucin in the shape of oily circular-looking discs
will be deposited.

Leucin obtained from urine is not crystalline,
but forms circular oily-looking discs (Fig. 15, *a*) which
float on the surface of water;
they generally have a somewhat
yellowish appearance, owing to
the colouring matter of the urine.
If this form of leucin be dissolved
in boiling alcohol, the solution on
cooling will deposit leucin in
crystalline plates. Leucin is deposited from its alcoholic solution

Fig. 15.—*a*, Leucin;
b, Tyrosin.

in white shining plates, greasy to the touch, lighter
than water, and much resembling cholesterin in
appearance; it is distinguished from that substance
by its insolubility in ether. It is slightly soluble
in cold water, and very soluble in boiling water,
soluble in 600 parts of cold absolute alcohol, very
insoluble in ether, melting point 170° C.; is decomposed by nitrous acid, leucic acid being formed and
nitrogen given off. Distilled with dilute sulphuric
acid and manganese peroxide, it yields valero-nitrile,

carbonic acid, and water. Fused with caustic potash
it is transformed into potassium valerate, hydrogen,
and ammonia.

127. **Tyrosin** $C_9H_{11}NO_3$ is invariably associated
with leucin. To obtain it from the urine ; precipitate
the colouring and extractive matters with basic
lead acetate, and filter; decompose the filtrate with
sulphydric acid and filter ; the clear filtrate is to be
concentrated, and, on cooling, crystals of tyrosin will
be deposited. The crystals form long prismatic needles,
which cluster together to form stellate groups ; sometimes, when obtained from urine, these groups are so
closely aggregated together as to form balls of spiculated needles (Fig. 15, *b*). The crystals are sparingly
soluble in cold water and alcohol ; soluble in boiling
water and in acid and alkaline solutions, the solubility
being increased by the presence of extractive matters ;
insoluble in ether.

Tyrosin treated with nitric acid turns an orange-
red colour, which becomes yellow on heating ; this
touched with a drop of liquor sodæ acquires a red tinge.

128. **Mucus and pus.**—Pure mucus forms a
clear translucent mass; mingled with it, however, are
epithelial cells derived from
various parts of the genito-
urinary tract (Fig. 16) and
pigmentary particles. Perfectly healthy urine always
contains a small quantity
of mucus, which, if the
urine be allowed to stand
in a glass vessel, will be
seen diffused as a fine cloud
in the lower stratum. The
epithelium is composed of

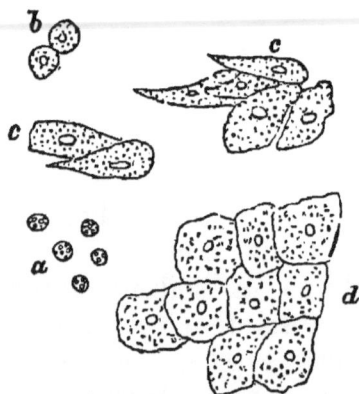

Fig. 16.—Epithelium in Urine.

bladder epithelium (Fig.
16, *d*), mixed, in the case of women, with the epithelium

from the vagina. When mucus is increased in quantity
by any morbid condition of the genito-urinary tract,
we have in addition mucus corpuscles, and epithelium
from the part of the tract affected ; as small rounded
renal cells (Fig. 16, *a*) and casts in nephritis; larger
rounded cells (Fig. 16, *b*) from the pelvis and kidney;
columnar cells (Fig. 16, *c*) from the ureter, or urethra ;
or large cancer cells in cancer of bladder, and granular
masses in tubercular disease. Pus is generally present in
the urine containing an excess of mucus ; its presence
is indicated by traces of albumin derived from the
liquor puris. To distinguish differentially between pus
and mucus, when both are present, is oftentimes somewhat difficult, especially if, in addition to the albumin
derived from the liquor puris, there is albuminuria
from kidney disease. The following reactions will aid
us in coming to a conclusion : (*a*) the addition of
liquor potassæ to a urine containing an excess of
pus renders it thick and gelatinous, whereas, if mucus
be in excess it is rendered thinner; (*b*) mercuric
chloride added to the urine, if pus is present, the
pyin is precipitated, but not mucin ; the precipitate
should be filtered off, and if mucin is present it will
be precipitated from the filtrate on the addition of
acetic acid ; (*c*) under the microscope it is difficult to
distinguish mucin corpuscles from pus corpuscles and
other young cell-forms, all swelling up and losing their
granular surface, whilst the nuclei become more visible
on the addition of acetic acid ; if, however, mucus is
in excess, a coagulation will often occur in the fluid
when a drop of acetic acid is added, from precipitation
of the mucin. The clinical character of the urine also
may afford a clue. When mucus is in excess, the urine
speedily undergoes acid fermentation and becomes
more acid, which speedily passes on to ammoniacal
fermentation. Purulent urine is usually neutral,
and undergoes alkaline fermentation slowly. Urines

containing an excess of mucus, if acid, frequently deposit the mucin in small shreds and cylinders, these
are soluble in liquor potassæ. On the other hand,
purulent urine, when alkaline, deposits white ropy
strings, which are soluble in acid.

Extraneous substances.—Fungi, entozoa, hair from
dermoid cysts, fœtal bones, etc., may all find their
way into the urine, and are best recognised by their
microscopic appearance. Sand, cane-sugar, mortar,
etc., may be added by hysterical patients for the purpose of deception.

129. **Detection of lead in urine.** — Many
organic and inorganic poisons are eliminated by the
kidneys and pass into the urine. In cases of poisoning by these substances, however, it is more usual to
examine the vomit and fæces than the urine, or if the
patient has died, to seek for them in the tissues. The
method to be employed in these cases is described
§ 151. It is convenient here to describe, however,
the process for the detection of lead in urine, as it
is frequently required for clinical purposes in cases of
plumbism, to watch the gradual elimination of lead
under treatment, especially after the administration of
potassium iodide. As the amount of lead is often
extremely minute, the whole twenty-four hours' urine
should be employed. This should be evaporated to a
treacly residue in a large porcelain dish, and this
transferred to a platinum or porcelain crucible, and
heat applied, gently at first, till nothing but a grey ash
is left. The soluble salts are then removed by the
addition of boiling distilled water, till a drop of the
washings gives no residue when evaporated on a
glass slide. The washings are to be examined with
sulphydric acid, to see that they contain no lead.
The residue is then treated with equal parts of
distilled water and nitric acid, and the mixture boiled
for a few minutes ; it is then filtered, and the filtrate

tested for lead : (1) By the addition of sulphydric acid, giving a black sulphide of lead ; (2) potassium iodide, a yellow precipitate of lead iodide ; (3) potassium chromate, a yellow precipitate of lead chromate, insoluble in dilute acids. These precipitates, together with the residue not dissolved by the nitric acid, and the precipitate given by the sulphydric acid, if any, with the first washings with hot distilled water, must be placed on a filter and dried. The dried residue mixed with sodium carbonate, and the mixture placed on charcoal and fused by the blow-pipe, when a bead of metallic lead will be obtained, which represents the whole of the lead present in a metallic state in the twenty-four hours' urine.

130. **Urinary calculi.** — (*See* Morbid products, chapter vi.)

CHAPTER V.

MORBID CONDITIONS OF THE DIGESTIVE SECRETIONS.

131. **Saliva,** as presented for clinical examination, is mixed with oral mucus, and oftentimes contains portions of food in a state more or less decomposed. It may be obtained fairly pure by directing the patient to wash his mouth out thoroughly with a dilute warm solution of bicarbonate of soda, and afterwards with cold spring-water. The inside of the mouth should then be lightly brushed with a glass rod moistened with a little dilute acid, when the mouth will fill with a considerable amount of clear viscid fluid. To obtain saliva absolutely pure, however, a small canula should be introduced into the ducts of the respective glands. The saliva obtained from the parotid and submaxillary glands

differs in quality; the former being rich in ptyalin
and poor in mucin, whilst the submaxillary gland
contains a considerable amount of mucin, and but
little ptyalin. The physiological effect of the
stimulation of the nerves supplying these two glands
has been thoroughly investigated and recorded in
physiological text-books. Here it will only be
necessary to recall the results. Thus, stimulation of
the parotid by Jacobson's nerve, leads to a watery secretion, containing little albumin, ptyalin, and salts. Irritation of the sympathetic causes no secretion, but irritation of both Jacobson's nerve and the sympathetic at
the same time, gives rise to an abundant secretion, in
which the organic constituents abound, but the salts
are only slightly increased. With the submaxillary
gland, stimulation of the chorda tympani produces an
increased flow of saliva ; if, however, the stimulation
be directed to the sympathetic filament exclusively,
the secretion, though abundant, becomes thicker and
more gelatinous than from stimulation of both filaments.
Section of the nerves without stimulation often leads
to the discharge, for days and weeks together, of a thin
aqueous saliva, the so-called paralytic saliva. Atropin,
as is well known, paralyses the action of the gland,
but other substances, such as pilocarpin and eserin,
stimulate the secretion. Langley has shown (*Journal
of Physiology*, vol. i.), that by injecting atropin into
the duct of the submaxillary, the secretion can be
stopped, but by injection of pilocarpin it is restored. The sublingual and buccal glands also
secrete saliva ; but in the human subject their
secretion cannot be obtained separately for examination, since it involves tying the ducts of the other
glands.

The mixed saliva has the following approximate
composition in 1,000 parts : Water 994·94 ; Solids 5·06
(ptyalin 1·2, mucin 1·3, fatty matters 1·1, salts 1·6).

The *ptyalin*, or diastatic ferment, may be separated
approximately pure by precipitating fresh saliva with
dilute normal phosphoric acid, and then adding lime-
water ; filter off precipitate, and dissolve it in distilled
water, from which it is to be precipitated by alcohol,
collected on a filter, washed repeatedly with a mixture
of alcohol and water, dried, and weighed. As ptyalin,
however, has never been obtained quite pure, its
chemical composition and reactions are doubtful,
except its energetic action on starch, which it converts
into maltose and glucose. *Mucin* may be obtained by
allowing saliva to fall into a beaker containing some
dilute acetic acid, when it will deposit in stringy
flakes ; these should be collected on a filter, and
washed with alcohol, dried, and weighed. Pure
saliva also contains traces of serum albumin and
serum globulin. The fatty matters are estimated
from the etherial residue as for blood (§ 93). The
inorganic residue is estimated (*a*) from the bases
by incineration (§ 101); (*b*) from the acids by volumetric
determination directly from the saliva, after the
removal of the organic constituents, as with the
estimation of phosphoric, hydrochloric, and sulphuric
acid in urine (§§ 113, 114, 115). The salts consist
chiefly of calcium phosphate, and carbonate, and are
deposited, unless care is taken, round the teeth as tartar;
sometimes they deposit in the salivary ducts, and form
a calculus. The saliva also contains *potassium sulpho-
cyanate*, which is of some interest, from a medico-legal
point of view, in opium poisoning (§§ 83, 136).

The variation in the quality and quantity of saliva
excreted under various morbid conditions has been very
inadequately studied. The following are the chief facts
of importance that have been ascertained. The average
daily amount excreted has been placed at 1,500 grms. ;
this is probably too high, and 800 to 900 grms. is nearer
the mark. The daily excretion is increased by dry

food. The excretion is diminished in pyrexia, and by the action of certain drugs, particularly belladonna. Claude Bernard found that dilute alcohol, dilute alkaline solutions, and saliva were all powerful excitants of the gastric secretion when introduced into the stomach by means of a gastric fistula; but of the three saliva was the most powerful. Saliva, then, must be regarded as having a twofold function ; its diastatic action, and as an exciter of gastric secretion. Diminution of its secretion must therefore lessen the digestive powers, both as regards the albuminous as well as the starchy constituents of the food. The saliva of new-born infants has no diastatic action on starch. The normal reaction of saliva is alkaline, it is often neutral, sometimes acid. The acidity in most cases is due to decomposition of organic substances in the mouth, though in some diseases, as diabetes, acute rheumatism, and mercurial salivation, the saliva is acid as it comes from the ducts of the glands. The diastatic action of saliva is best manifested in dilute alkaline solutions at 40° C., but it also acts in neutral and very dilute acid solutions ; strong alkalies and acids, and temperatures above 70° C., stop its action entirely. In conditions of debility, and under the influence of certain drugs (sialogues), of which mercury, pilocarpin, and eserin are the chief, the amount dis charged in the twenty-four hours is greatly increased. In water brash the aqueous discharge from the mouth undoubtedly proceeds in great measure from the salivary glands. Salivation is often met with in hysterical females, and is sometimes of an intermittent character. I have noticed profuse salivation after injury to the upper part of the food tube in a case of attempted sulphuric acid poisoning ; the sialorrhœa came on some time after the separation of the eschars and healing of the ulcers. Physiological remedies were tried without avail, and the salivation continued till the patient's

general health was restored, when it gradually subsided.
Altered saliva is often retched up from the stomach
in cases of chronic gastric catarrh, especially that form
dependent on alcoholism. This retching is most
frequent in the morning, and stringy masses are
brought up. These consist of mucin derived from
the saliva swallowed during sleep, and deposited from
it by the action of the acetic and other acids; produced
by fermentative action going on in the stomach, just
in the same way as mucin is thrown down from saliva
when dropped into a vessel containing dilute acetic
acid.

Salivation, as is well known, follows the continued
use of mercury ; and extremely minute doses will, with
some people, rapidly induce the condition. In most
instances we can trace the salivation to the use of the
drug ; but in some cases, when it has been accidentally
introduced into the system, either by the stomach
in quack medicines, or through the skin by handling
some material the deleterious nature of which was
unknown, it becomes necessary to examine the saliva
in order to discover the presence of mercury, and so
get a clue to the cause of the salivation.

132. **Detection of mercury in saliva.**—For
this purpose the saliva is to be collected for twenty-four
hours, and dilute hydrochloric acid (one part of acid
to nine parts of water) added to it. The mixture is
heated for two hours in a water-bath, filtered, and the
filtrate, which should be labelled *a*, concentrated to
half its bulk. The precipitated matters in the filter
are then placed in a beaker, filled three parts full
with dilute hydrochloric acid (one part acid to six
water), and the whole heated over a water-bath,
adding from time to time small quantities of potassium
chlorate, and constantly stirring to dissolve the
organic residue. When this is completely dissolved,
filter, and add filtrate to the previous filtrate *a*.

M

Concentrate the mixed filtrates to one-fourth their bulk. The solution contains any mercury that may be present as bichloride. To prove the presence of mercury, (1) place a drop of the solution on a gold or copper coin, and touch with blade of knife; a bright silvery stain will appear. (2) Place a few strips of *pure* copper-foil in a test-tube, and add a little of the solution, and boil; the mercury will be deposited on the surface of the copper-foil. Remove the strips and wash them with very dilute solution of ammonia, and dry them between blotting-paper. Then place them at the bottom of a narrow glass tube (German glass), and apply heat; the mercury will be volatilised, and deposited as a ring of minute globules at the upper end of the tube. The character of these globules can generally be recognised by the eye. If, however, they are too small, remove the strips of copper from the tube, and dissolve the ring by the addition of a drop or so of dilute nitro-muriatic acid, and gently evaporate the solution. Dissolve the residue in a little water, and divide into two equal portions; (*a*) tested with a drop of dilute solution of potassium iodide, it gives a red precipitate of mercuric iodide, soluble in excess of potassium iodide solution; (*b*) a drop added to solution of caustic potash gives a yellow precipitate of hydrated mercuric oxide, insoluble in excess of liquor potassæ. (This process is available for the detection of mercury in the solid tissues, care being taken to divide them very finely before treating with hydrochloric acid and potassium chlorate.)

133. **Gastric juice** is generally obtained for examination from a gastric fistula in animals, and contains saliva, unless the precaution of previously tying the ducts of the salivary glands has been adopted. The following is an approximate analysis of gastric juice obtained from the human subject, and

which, of course, contains a certain amount of saliva : Water, 994·6; Solids, 5·4 ; pepsin 3·02, free hydro-chloric acid 0·22, alkaline chlorides 2·0, phosphates of lime, magnesia, and iron, 0·15.

134. **Reaction of the gastric juice** is acid, and varies considerably, according to the statements of different observers. Thus, Richet, from numerous observations on a patient after gastrotomy, gives the average acidity as 1·7, with a maximum of 3·4 and a minimum of 0·5 per thousand ; Schroder, from obser-vations made in a female with gastric fistula, records it as low as 0·2 ; Schmidt, from experiments on dogs, gives an average of 2·5, and Szabo, with the same animals, 3· per thousand. These variations need not be considered contradictory, the acidity of the gastric juice no doubt depending much on the nature of the physiological stimulus that excites it. This supposi-tion receives support from the observations of Schmidt, who found the juice of herbivorous animals had a lower degree of acidity than that from carnivorous animals. One point, however, is certain, that the acid is present in a very dilute state, thus confirming the results obtained by experiments with artificial gastric juice, in which a degree of acidity of 0·2 per cent. of hydrochloric acid is found to be most effective. With regard to the amount of acid withdrawn from the blood by the gastric secretion during the twenty-four hours, it is impossible to speak with any certainty, since the quantity of gastric juice secreted during that period has never been definitely ascertained. Grüne-wald, in a case examined by him, states it as twenty-three imperial pints, but this was undoubtedly under pathological conditions. Parkes considers if we put it at twelve pints we shall be within the mark. Lehmann, drawing conclusions from experiments on animals, concludes that the secretion of gastric juice in the twenty-four hours amounts to one-tenth of the

whole weight of their body. This, per man, would
represent something like 14 lbs. avoirdupois. Unfortunately, as Richet well observes, the data upon
which these calculations are founded are very uncertain, since it is extremely difficult to determine the
relative proportion of the true gastric secretion
from the mucus mixed with it, and also to make
allowance for what passes off accidentally during
the experiment by the pylorus, and what is absorbed
by the veins of the stomach. Moreover, even if
these obstacles should be overcome, the intermittent
nature of the secretion would make it difficult
to arrive at very definite conclusions. It has now
been incontestably proved that the acidity of the
gastric juice is due to free hydrochloric acid. Richet
has shown that in the fresh secretion this is the only
mineral acid present. Lactic, acetic, and butyric
acids are also met with in gastric juice, the result of
fermentative changes occurring in the stomach. In
certain morbid conditions they may be considerably
in excess of the hydrochloric acid (indeed, that acid
may be very scantily secreted), and thus, by causing
delay in gastric digestion, lead to the formation of
these organic acids. In many cases of acid dyspepsia
it is a matter of importance to determine whether
the acid in the vomited matters contains a due proportion of hydrochloric acid, or whether the organic
acids are in excess. In the former case, the acidity
will arise from hyper-secretion ; in the latter, from
fermentative changes ; or both acids may be in excess.
Till recently we had no ready means of determining
the nature of the acid present in the vomited matters,
and therefore uncertainty frequently existed as to the
conditions under which the acids expelled were formed
in the stomach. Richet, however, has suggested a
method by which the nature of the acid can be
accurately determined, and has thus supplied us with

an additional means of diagnosis in those cases of
stomach disease attended with the vomiting of acid
matters. His method is based on the fact that if an
aqueous acid solution be shaken with ether the latter
removes a constant quantity of the acid. This, in
the case of mineral acids, is extremely small, but with
organic acids the removal is considerable. The specific
ratio which exists, after an aqueous solution of acid
has been agitated with ether, between the quantity of
acid taken up by a certain volume of ether and that
which remains in an equal volume of the solution
after it has been treated with ether, is called the " co-
efficient of partage," a term originally applied by
Berthelot. The co-efficient of partage, in the case of
mineral acid, is high (above 500) because the
quantity of acid yielded to the ether is small ; the co-
efficient for the organic acids is low, for the opposite
reason. The following example will render the matter
clearer : 100 grammes of water containing 11 grammes
of lactic acid, and 100 grammes of ether agitated
with this solution removes 1 gramme of acid ; so
when we determine the acidity of the two fluids we
find that of the water to be 10 and that of the ether
1. But supposing the degree of dilution to be ten
times greater than in the first case, then 100 grammes
of water which contain $1 \cdot 1$ grammes of lactic acid,
agitated with an equal weight of ether, will yield to
the ether $0 \cdot 1$ gramme, and retain 1 gramme, the co-
efficient of lactic acid is therefore said to be 10. The
co-efficients of many other organic acids have been
determined. Some of the most important, as having a
bearing on animal chemistry, are succinic acid $c'=6$,
benzoic acid $c'=1 \cdot 8$, oxalic acid $c'=9 \cdot 5$, acetic acid
$c'=1 \cdot 4$. So far as concerns one acid in solution the
operation is simple enough ; but when we have to deal
with a mixture of two or more we must have recourse
to a series of agitations with ether, so that we may

separate the acid which is the most readily soluble in
ether from the one that is less so. By such repeated
treatment of the original acid solution with ether, and
recording the co-efficient of partage after each operation, we are able to obtain the true co-efficient of
partage for each acid. Richet has found by repeated
experiments that gastric juice, when freshly taken
from a fistula, has a high co-efficient of partage corresponding to that of hydrochloric acid, whilst by
keeping the juice some time the co-efficient of partage
gradually fell, denoting the increase of the organic
acids from fermentative changes taking place in the
secretion. Another argument pointing conclusively
in favour of the acidity of the gastric juice being due to
hydrochloric acid, is that if we estimate the bases and the
chlorine separately, we find that there is more chlorine
than is required to convert all the bases into chlorides.
As Ewald remarks, the excess of chlorine can only
exist as free hydrochloric acid, or in an organic
combination. In addition to this, the fact that the
determined acidity of the gastric juice as obtained
experimentally by means of fistulæ closely corresponds
with the degree of acidity at which the artificial
gastric juice is most active (viz., 0·2 per cent. of
hydrochloric acid), is another point proving that
hydrochloric acid is the acid concerned in gastric
digestion, since it has been shown experimentally that
when artificial gastric juice is prepared with lactic
acid instead of hydrochloric acid, a degree of acidity
six times greater than the natural acidity of the
gastric juice is required to effect digestion, whilst if
acetic acid is employed the degree of acidity required
is twice as great.

We must now proceed to consider the manner in
which the hydrochloric acid of the gastric juice is
separated in a free state from the alkaline blood. It
is only recently that an explanation has been offered

to account for this seeming paradox. In 1874,* in order to elucidate this point, I made in the laboratory of the Charing Cross hospital a series of experiments, in which I found, by introducing an alkaline solution consisting of sodium bicarbonate (5 per cent.) and neutral sodium phosphate (5 per cent.) into a small U-tube, fitted with a diaphragm at the bend, and passed a weak electric current through the solution, that in a short time the fluid in the limb connected with the negative pole increased in alkalinity, whilst the fluid in the limb connected with the positive pole became acid from the formation of acid sodium phosphate. Now, one of the chief salts in the blood is undoubtedly sodium or potassium bicarbonate, an acid salt with an alkaline reaction; and neutral sodium phosphate has also an alkaline reaction. The decomposition which occurs between them may be represented as follows :

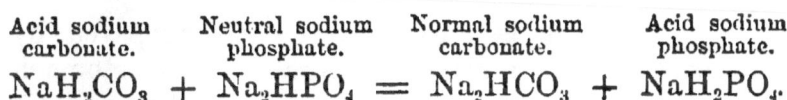

Acid sodium carbonate.	Neutral sodium phosphate.	Normal sodium carbonate.	Acid sodium phosphate.

$$NaH_2CO_3 + Na_2HPO_4 = Na_2HCO_3 + NaH_2PO_4.$$

The above reaction explains the presence of acid sodium phosphate in urine. To account for the formation of free hydrochloric acid in the gastric juice, sodium chloride is substituted for the neutral sodium phosphate, the decomposition in this case being

Acid sodium carbonate.	Sodium chloride.	Normal sodium carbonate.	Hydrochloric acid.

$$NaH_2CO_3 + NaCl = Na_2HCO_3 + HCl.$$

Maly, however, who subsequently investigated (1877) the subject with great care, has come to the conclusion that the hydrochloric acid is derived from

* *Lancet*, p. 29, July 4th, 1874.

the decomposition of neutral sodium phosphate with calcium chloride, as described §§ 79—84.

Practically, it matters little which view we adopt, since all the salts named are present in the blood ; the important fact being, that out of the body a weak electrical current will separate the acid from its base. Whether the decomposition occurring in the body is due to the same agency must for the present remain a matter of conjecture. Whatever be the nature of the agency that causes the decomposition, it must be a powerful one to effect the separation of hydrochloric acid from bases for which it has such a strong affinity as soda or lime. The decomposition, however, once effected in the blood, there is no difficulty in explaining the presence of free hydrochloric acid in the stomach, since Graham showed many years ago that this acid possesses high diffusive power, and passes from a mixture through a dialyser with great rapidity.

135. **Pepsin** is most conveniently prepared by dissecting off the mucous membrane of the stomach of a recently killed animal, rejecting the pyloric extremity. The mucous membrane is then sliced into thin shreds, and macerated in dilute phosphoric acid till dissolved. The mixture is then strained through coarse muslin, and the filtrate precipitated with an equal volume of lime-water. The precipitate is collected in a filter, well washed, and dissolved in dilute hydrochloric acid. The solution is then placed in a glass flask, and a saturated solution of cholesterin, in one part of ether and four parts of alcohol, is passed down to the bottom of the flask by means of a tube, and the whole mixture well agitated. The cholesterin separates, mingling with the pepsin. The cholesterin is removed by repeated treatment with ether, leaving the pepsin as a greyish-white powder, insoluble in water, alcohol, and ether, but very soluble

in dilute acids. The acid solutions are precipitated by alcohol, and by neutral and basic lead acetate, but not by strong nitric acid, tannic acid, or mercuric chloride. Pepsin by itself has no action on albuminous substances, but, in conjunction with dilute acid, converts them into peptones. By this conversion albuminous substances become more diffusible. Thus, taking the diffusion rate of ordinary albumin at 100, we find that, when converted into peptone, the diffusion rate is 7·1—9·9; in other words, it is increased about twelve times. The natural acid of the gastric juice is, as stated in the preceding paragraph, hydrochloric acid, and the degree of acidity at which digestion is best effected was put at about 0·2 per cent. of the real acid. The proportion between the quantity of acid and the quantity of pepsin required to reduce a given amount of albuminous material may be thus stated. If, during artificial digestion, the process comes to a stop, the addition of some dilute acid will set it going again. After a time, however, the addition of acid is not sufficient, and more pepsin has to be added. The quantity, however, of pepsin required, as compared with the quantity of albumin digested, is really very small. The action of gastric juice on the various food-stuffs can be readily studied by means of a glycerin extract of pig's stomach,* and a 0·2 per cent. solution of HCl. With *fats*, the gastric juice dissolves the albuminous envelopes of the fat cells, whilst the temperature (40° C.) at which digestion is carried on renders the solid fats fluid, but no real chemical change occurs. Gastric juice has no action on

---

* This is made by mincing the mucous membrane of a pig's stomach, and covering it with glycerin. After standing forty-eight hours, the glycerin is strained off, and more glycerin added to the residue, and this again strained off. The process is repeated till all the pepsin has been extracted.

the *starchy* matters of the food, but it does not put a stop to the conversion of starch into glucose by the saliva. Gastric juice dissolves *gelatiniferous* tissues ; that is, gelatin, after digestion with artificial gastric juice, loses the power of gelatinising. The casein of *milk* is coagulated before being converted into peptone ; this curdling is apparently caused by some ferment which sets up lactic acid fermentation (Hammarsten) of the milk sugar, and is not due to the acid of the gastric juice itself. All *proteid* bodies, with the exception of lardacein, are converted into peptones by the action of pepsin in dilute acid solutions. The circumstances chiefly influencing gastric digestion may be thus enumerated : Digestion proceeds best at temperatures between 35° and 40° C. ; boiling destroys all action ; neutralisation, or the presence of too much acid, arrests the action ; the concentration of the products of digestion retard digestion ; minute subdivision, by increasing the surface acted on by the gastric juice, favours digestion ; alcohol, strong alkaline mixtures, and salts of the heavy metals, also interfere with the process.

135 *a*. **Peptones,** as obtained by the action of gastric juice on proteid substances, are white amorphous bodies, giving an acid reaction to litmus paper, soluble in water. Their solutions are not coagulable by heat, nor by mercuric salts, tannic acid, mineral acids, or alkalies. Bile, added to an acid solution of peptone, precipitates it. The reason of this is not clear ; it is probably analogous to the precipitation of parapeptone that occurs on neutralisation with sodium carbonate. When quite pure, *i.e.*, free from undigested albumin, they should give no reaction with Millon's test or the xantho-proteic reaction. Mixed with Fehling's solution, they give a purple-violet ; but, when floated on the surface, the coloration at

the junction of the two fluids is rosy red. According to Dr. George Johnson, they are precipitated by picric acid, the precipitate being re-dissolved when heated. This last reaction, in my opinion, is due to the presence of a bye-product, parapeptone or hemi-albumose, rather than characteristic of the true peptone. The peptones all possess lævo-rotatory power. There are several varieties of peptones. According to the original view of Meissner, there is a bye-product resembling syntonin, which he called parapeptone; intermediate products, called metapeptone and A and B peptones; and the true, or c, peptone.. Kühne considers that the initial step in both gastric and pancreatic digestion is the breaking up of albumin into *anti-albumose* and *hemi-albumose*. The former is converted into *anti-peptone*, which undergoes no further change, either by the action of gastric or pancreatic digestion, and gives the reactions above described as characteristic of true peptone ; the latter is converted into *hemi-peptone*, which, by the action of *trypsin*, the pancreatic ferment, is converted into leucin and tyrosin, and to which we shall refer when speaking of the pancreatic juice. With regard to the two bye-products, *anti-albumose* and *hemi-albumose*, the former may be regarded as identical with Meissner's parapeptone, and the latter with his A peptone. With regard to their characters, *anti-albumose*, or parapeptone, is insoluble in water, soluble in dilute acids ; and the acid solution is precipitated by potassium ferrocyanide, mercuric chloride, tannic acid, and picric acid. Hemi-albumose, or Meissner's A peptone, is soluble in water at 70° C., but is re-precipitated on cooling, soluble in 10 per cent. solutions of sodium chloride ; it thus resembles the "albumin" found in the urine of a case of osteo-malacia by Dr. Bence Jones. With regard to the composition of these bodies, they are generally regarded as *hydrates of*

albuminate, and are formed by albumin taking up
water, as starch does to form glucose. Henninger is
said to have proved this by changing peptone back
to syntonin by the abstraction of water. The clinical
interest attaching to the peptones is their appearance
in urine under various morbid conditions. The
method for isolating these bodies is described § 118,
page 148.

136. **Vomited matters.**—Vomiting is induced
(*a*) directly, by stimulation of the stomach; (*b*) indirectly, through the irritation of other and distant
organs. In the first category we have to consider the
vomiting in relation to disease of the organ, cancer,
ulceration, stricture of the pylorus, or poisoning by
corrosive poisons, etc.; and here we have to determine
the conditions with regard to the digestion of the
food ingested, the acidity of the gastric secretion, the
presence or absence of blood or bile, cancer cells, and
sarcinæ. In the second category, viz., vomiting produced by irritation of distant parts, the retching
produced by tickling the fauces is the most familiar;
but reflex vomiting may be produced by irritation of
any organ; thus, the vomiting in uterine disease on
the passage of renal or biliary calculi; the early
vomiting of phthisis and in disease of the brain; or
it may be caused through the nervous system by toxic
agents, as tobacco, lobelia, opium, etc. ; or by blood
poisons, as in erysipelas, septicæmia, gout, etc. For
clinical purposes, the chemical examination of vomited
matters may be limited (1) to an inquiry into the
nature of acid present, whether due to hyper-secretion
of gastric juice (excess of hydrochloric acid), or from
fermentative changes occurring in the stomach (excess
of lactic or acetic acid) ; (2) the detection of poisonous
substances in the vomit.

(1) *Determining amount and nature of acid
present in vomit.*—In directing our treatment with

regard to stomach affections it is of the utmost importance for us to recognise whether the acid present in the vomited matter is due to hyper-secretion of gastric juice (hydrochloric acid) or to fermentative changes, lactic acid, etc.* Nor is it sufficient to rest content with one examination only, since in these cases it often happens that both forms may be present, but that one preponderates more at one time than another. Thus, Dr. Golding Bird states that in a case of scirrhous pylorus he found at one time a quantity of free hydrochloric acid in one pint of vomit equal to twenty-two grains of the pharmaceutical acid, with an organic acid sufficient in quantity to neutralise seven grains of pure potash. At another time the hydrochloric acid had nearly disappeared, and the quantity of organic acid in each pint required for saturation nearly seventeen grains of the alkali. If, therefore, we are desirous of giving relief we ought to follow regularly the variations in the character of the acid in the vomited matters. For this purpose the following plan of procedure will be found easy as well as reliable. The vomit is poured into a tall cylindrical glass jar, and allowed to settle. Of the supernatant fluid draw off 100 cc., or if so much cannot be obtained, then 50 cc. or 25 cc. Shake this up with an equal weight of ether in a cylindrical tube, set aside till the ether has separated from the mixture. Then remove etherial solution by means of a pipette. Now, as it has been stated (§ 134, page 181), that the organic acids are more readily removable by ether than the mineral acids, so that the acidity of the etherial solution represents in the main the acidity due to organic acids, whilst the

---

* I have endeavoured to discriminate between the forms of "acidity" met with in functional disorders of the stomach by reference to the condition of the urine. (*See* chaps. ii. and iii., "Morbid Conditions of the Urine." Churchill. 1882.)

acidity of the vomit, after extraction by ether, represents nearly the whole of the mineral acids (hydrochloric). If, therefore, we take the acidity of the
supernatant fluid of the vomit before agitation with
ether, by the same process as directed for determining
the acidity of urine (§ 107, page 112), and again afterwards, we can judge by the difference in the acidity
whether the organic or mineral acid is in excess.
If the former, the reduction in the acidity will be
marked ; if the latter, it will not be considerable. If
we desire to find out what organic acids are present,
the vomit will have to be treated repeatedly and successively with ether till the co-efficient of partage of
each has been determined. There is no necessity, however, for engaging on this lengthy process ; for clinical
purposes it is sufficient to determine whether we are
dealing in any given case with excess of organic or
mineral (hydrochloric) acid.

(2) *Detection of poisons in vomit.* — Without
engaging in the elaborate and exact processes of
analysis requisite for toxicological purposes, and
which, when required for medico-legal purposes, should
always be conducted by a professional expert, medical
men frequently have to decide, when called to a case
of urgent vomiting, whether it is due to a poisonous
agent. In these cases there is generally no time to
refer the matter to an analytical chemist. Every
medical man, therefore, ought to be able to conduct a
preliminary enquiry, so as to gain some insight as to
the nature of the case with which he is dealing. As
a · rule a clue is afforded by the character of the
symptoms, which show whether the poison belongs to
the irritant class, or is a corrosive substance, or one of
the vegeto-alkaloids ; and we have generally some of
the substance taken left, which also materially assists ;
thus we readily recognise oil of almonds, vermin paste,
oil of vitriol, oxalic acid, and the like. But where the

substance cannot be obtained for inspection, and the symptoms are obscure, we are forced to make an elementary analysis. The plan to be adopted is as follows:—Reserve a sufficient portion of the vomit for detailed examination at the hands of an expert; place this in a strong bottle tightly corked and sealed. With regard to the surplus proceed according to the following method, which deals with the most common forms of poisoning.

(*a*) *Volatile poisons.*—Place a small quantity of the fluid in a test tube, and gently warm the fluid; if prussic acid is present, the peculiar odour will be evolved. The best confirmatory test for this is to place a little of the vomit on a small watch-glass or glass slide, and add, by means of a stirring-rod, five or six drops of strong sulphuric acid; hold over the watch-glass another moistened with liquor potassæ. The fumes of the hydrocyanic acid form potassium cyanide. This is tested by stirring it with a rod dipped successively in a solution of a ferrous salt, ferric salt, and hydrochloric acid, when, if *hydrocyanic* acid is present, Prussian blue will be developed. If no hydrocyanic acid be found, apply strong heat to another portion of the vomit placed in a narrow tube, and carry the tube into a darkened room to see if fumes of *phosphorus* are given off.

(*b*) *Corrosive poisons.*—Test with litmus paper; if strongly acid, probably oxalic acid (§ 47), sulphuric acid (§ 82), hydrochloric acid (§ 79), or carbolic acid (§ 50); if strongly alkaline, due to some of the caustic alkalies or alkaline salts.

(*c*) *Metallic irritant poisons.*—Some of the vomit, acidulated with pure hydrochloric acid, is placed in a test-tube, and a strip of perfectly clean copper, dipped in alcohol to prevent fatty matters adhering to it, is then immersed in the acidulated mixture, and heat

applied for about twenty minutes. If arsenic, antimony, or mercury is present, the copper will be
stained black. Remove the copper strip, and wash it
with a little very dilute solution of ammonia, and
then dry it between folds of blotting-paper. Then
place in a narrow glass tube, and proceed as directed
for the detection of *mercury* (§ 132), *arsenic* and
antimony (§ 151).

Strychnine and morphia.—A small portion of the
vomit is to be rendered strongly alkaline with sodium
carbonate, and agitated with four times its volume of
ether. After the etherial solution has formed a layer
on the surface, it is to be removed by means of a
pipette, and allowed to evaporate spontaneously in a
watch-glass ; the residue is to be examined for morphia (§ 70), strychnia (§ 71). It is often sufficient
to place a drop of the etherial solution on the tongue,
by means of a glass rod, to learn the character of
the alkaloid. If a frog can be obtained, evidence
of poisoning by an alkaloid is then most readily
obtained by injecting some of the ether extract under
the skin.

Opium.—Evaporate a small portion of the vomit
on a porcelain dish ; touch the dry residue with ferric
chloride ; a cherry-red colour, which does not disappear
when touched with mercuric chloride, indicates the
presence of meconic acid (§ 83).

137. **Gases in the stomach.**—The nature of
the gaseous contents of the stomach in health and
disease will be best considered with reference to the
gases in the intestinal canal (§ 150).

138. **Bile.**—The composition of the bile varies
considerably, the proportion of its solid constituents
ranging from 9 to 17 per cent., being always most
after a meal. The following analyses, (1) made by
Frerichs, is from the bile taken from the gall-bladder
of a healthy man killed by an injury ; (2) the mean of

five analyses by Hoppe Seyler, obtained from bodies in the post-mortem room :—

	No. 1. Frerichs.	No. 2. Hoppe Seyler.
Water	85·92	91·68 ⎱
Inorganic salts	·78	
Organic matter	13·30	8·32 ⎰
Mucus pigment	2·98	1·29
Bile salts	9·14	3·90
Fat	·92	0·73
Soaps ⎫		⎧ 1·39
Cholesterin ⎬	·26	⎨ 0·35
Lecithin ⎭		⎩ 0·53

Obtained fresh, as it flows from the liver, it is a thin, transparent fluid, of golden-yellow colour, like yolk of egg, of a very bitter taste, of alkaline re-action, and an average sp. gr. of 1·018. When obtained after death, the colour is of brownish-yellow; it acquires a tenacious consistence from the presence of mucin, which is furnished by the gall-ducts and gall-bladder. Bile mixes freely with oil and fat, and, when shaken with them, forms an emulsion, which renders their passage through animal membranes more easy. Added to a solution of gastric peptones, a precipitate occurs; this precipitate is caused, no doubt, by the alkaline salts of the bile precipitating the parapeptone from its acid solution. The quantity of bile secreted increases suddenly after a meal, reaches its maximum in about two hours, and then gradually declines. The quantity of bile dis-charged daily may be put at forty ounces. Thus, Carpenter, from the experiments of Nasse, Platner, and Stackman, calculated that a man weighing 154 pounds should secrete this quantity; whilst Murchison

N

records (Case CLXXII., *Diseases of Liver*) an instance
where quite two pints of bile were discharged daily
through a fistulous opening in the gall-bladder. Of
this quantity, only a small portion escapes by the
bowel, the remainder being re-absorbed in the intestines. Thus, the experiments of Bidder and Schmidt
on dogs have shown that only $\frac{1}{15}$th of the sulphur
originally passed into the intestine with the bile
appears in the fæces. Bischoff, again, has calculated
that about 46 grms. of the altered biliary acids are
discharged by man daily with his fæces, whilst Voit
has shown that the average daily quantity formed by
the liver amounts to 170 grms. ; therefore 124 grms.
must be otherwise disposed of. The observations of
Jaffé and McMunn have shown that the bilirubin of
the bile pigment is oxydised in the intestine to urobilin, in which condition it is absorbed, and passes off
from the body by the kidneys as the chief colouring
matter of the urine, either as urobilin or its more
oxydised product choletelin, only a portion of the
biliary pigment being discharged by the intestines
with the fæces. As Murchison pointed out, this
re-absorption of bile is, in fact, merely part of that
chemical circulation which is constantly taking place
between the fluid contents of the bowel and the blood,
the existence of which has been already alluded to
(§ 9). Bile has no action upon the digestion of proteid substances, beyond the negative effect of precipitating parapeptone, as noticed above. The bile
of some animals has an action on starch, converting it
into glucose ; and some observers state that, to a
slight extent, a similar action occurs with quite
fresh human bile. It acts, however, on the fatty
substances by forming an emulsion with them ; this
can be seen by placing a drop of cod-liver oil on a
glass slide, and adding a drop of fresh bile, when a
milky emulsion will be formed. Owing to the alkaline

salts, bile is capable of forming soaps with fatty
acids. Experimentally it has been shown, after applying a ligature to the common duct, that an
animal absorbed less fat than before. It has also
been pointed out that, where biliary fistulæ have been
established for a considerable time, nutrition was
only kept up so long as the loss was compensated by
increased food. Bile acts as a natural purgative, by
stimulating peristalsis ; and, since it possesses powerful
antiseptic properties, it arrests putrefactive fermentation in the intestines. A defective secretion of bile,
therefore, is one of the chief causes of flatulent
dyspepsia. In my work on " Morbid Conditions of
the Urine associated with Derangements of Digestion" (p. 49), I have pointed out that the form of
dyspepsia arising from deficient secretion of bile is
frequently attended with an alkaline condition of
urine, since the bile, being the chief secretion by
which the alkaline salts are discharged from the
blood, any hindrance, therefore, to the discharge of
the bile leads to their being eliminated in greater
quantity by the kidney. It is in this form of
dyspepsia that the greatest good is obtained by the
administration of dilute hydrochloric acid. Dr. Wickham Legg is of opinion that the passage of bile into
the intestine appears in some way necessary to the
formation of glycogen by the liver, since, after ligature of the bile-duct of a cat, the diabetic puncture
failed to give rise to sugar in the urine. In jaundice,
a yellow tingeing of the skin, as well as the other
tissues and fluids, takes place, owing to the presence
of biliary matters in the circulation. Jaundice is generally considered as being either *heptogenous*, caused
by re-absorption of bile either from obstruction of the
bile-ducts or from disturbances of the portal circulation, and *hæmatogenous*, from changes occurring in
the blood. Among the causes leading to *heptogenous*

jaundice are, (1) simple catarrh of the bile - ducts, either by a mucus plug in the gall duct or by swelling of the mucous fold over the orifice of the duodenum, accompanied with gastro-intestinal catarrh, as is seen in malarial fever, secondary syphilis, pyæmia, etc. ; (2) direct obstruction of the ducts, as by impaction of gall stones or the presence of tumours ; (3) sudden changes of blood pressure, as hæmorrhage from the roots of portal veins, which favours the imbibition into the circulation of the secreted bile. The icterus menstrualis, and the jaundice so often observed in pneumonia of the right lung, are probably caused by the alteration of the blood pressure in the portal system.

Hæmatogenous jaundice is accounted for : (a) That when the destruction of liver tissue is extensive, the pigments of the bile are formed in the blood. (b) That the bile pigment in these cases is obtained by the dissolution of the blood corpuscles, and by the conversion of the hæmoglobin into bile pigment. Among the forms of jaundice, of hæmatogenous origin, are the jaundice of acute yellow atrophy, phosphorus poisoning, typhus, pyæmia, and septicæmia, and the jaundice following the bites of venomous animals. With regard to the theories supported in advance of a true hæmatogenous jaundice, it will be sufficient to say, with regard to the view that jaundice arises when the liver is extensively destroyed, by the accumulation of the bile pigment in the blood, that physiology has long since disposed of the view that the biliary elements are preformed in the blood ; they are only elaborated by the liver. With respect to the statement that the jaundice in these cases is due to dissolution of the blood corpuscles and the conversion of the hæmoglobin into bile pigments, experiment has shown that a variety of different substances, possessing quite a diverse character, introduced into the circulation, such as water, bile acids, ether, chloroform, etc., have the power of causing

bile pigment to appear in the urine; but that is quite
a different matter to converting hæmoglobin into bile
pigment in the circulation without the intervention of
the liver. If dissolution of the blood corpuscles, and
the conversion of the hæmoglobin into bile pigment
was, as the propounders of this theory maintain, the
cause of hæmatogenous jaundice, then scurvy, in which
the dissolution of the blood corpuscles is carried to
a great extent, would be invariably associated with
jaundice, which, however, is rarely the case. In that
obscure disease hæmatinuria (§ 120, page 161), where
the colouring matter of the blood appears in the urine
freed from the corpuscles, jaundice, though generally
present, is only slight, certainly not what would be
expected with such extensive destruction of blood
corpuscles. Indeed, it is probable that hæmatinuria
depends rather on some functional disturbance of the
liver, by which the effete hæmoglobin is not reduced to hæmatin, and from hæmatin to bilirubin,
as a failure of a normal process, than from any
positive evidence that the dissolution of the blood
corpuscles is caused by the introduction of a toxic
agent into the blood. Moreover, in addition to the
objections above urged to the existence of an hæmatogenous jaundice, the opinion is gaining ground that
in the cases where it is said to occur there exists some
overlooked catarrh in the capillary system of gall ducts,
and so that the jaundice is really heptogenous. The
microscopic examination of the livers of dogs poisoned
with phosphorus has shown incontestably that the finer
ducts were plugged with a colourless mucus; and when
we reflect how readily the bile has been shown, experimentally, to pass into the circulation on the slightest
increase of excretory pressure, we can conceive how a
very slight exudation in the finer tubes may cause very
appreciable jaundice, although the obstruction of the
tubes may not be visible to the naked eye. Lastly, the

supporters of the hæmatogenous theory have long-held
the view that in these cases the bile acids are not
found in the urine, yet in the jaundice associated with
pyæmia and septicæmia, which is one of their strongest
instances of a jaundice arising from blood poisoning,
the presence of the biliary acids in the urine has been
repeatedly demonstrated.   The position I would take
with regard to the existence of a hæmatogenous
jaundice is this, that there is a form of jaundice arising
from a primary morbid condition of the blood, which
leads to catarrh of the finer hepatic ducts, and so
causes jaundice from re-absorption of bile already
formed by the liver cells, and that the jaundice is not
occasioned by the conversion of hæmoglobin into bile
pigment in the circulation.   With this proviso it
is convenient to retain the term hæmatogenous as
distinguishing between the forms of jaundice arising
from primary changes, occurring in the condition of
the blood, leading to catarrh of the finer biliary ducts,
as apart from the jaundice arising directly (heptogenous) from obstruction.

The analysis of bile is conducted as follows.   The
specific gravity is taken by means of the specific
gravity bottle (§ 92), and the degree of alkalinity of
the secretion by the process, and with the same
standard solution as described in determining the
alkalinity of blood (§ 93).

139. The **mucin** is then to be removed by precipitation with alcohol or acetic acid.   On the addition
of either of these to bile, stringy ropes of mucin are
formed.   These must be allowed to subside, and the bile
thus clarified decanted off and passed through a fine
muslin filter.   This is divided into two portions (a)
evaporated at a gentle heat to one-fourth its bulk.
This forms *inspissated bile extract*.   (b) Evaporated to
half its bulk, mixed with an equal quantity of animal
charcoal, and introduced into a flask containing twice

its bulk of alcohol, it is digested till wanted. This forms *alcoholic bile extract*. The stringy ropes of mucin are then collected in a flask, and shaken up with ether to free them from fatty matters, and afterwards washed with water. When quite pure from adhering impurities, dissolve in a sufficiency of lime or baryta water, filter several times through animal charcoal to remove any colouring matter of the bile. Tests for mucin, § 26.

140. **Bile pigments** are obtained for examination from the inspissated bile extract or from gall stones. The chief pigment, *bilirubin*, can be obtained from either of the above-named by extracting successively with water, alcohol, dilute hydrochloric acid, boiling alcohol, and ether. Then boil the dry residue with pure chloroform, and distil the chloroform extract to near dryness, and then add several volumes of absolute alcohol; set aside for twenty-four hours, when an orange-red powder mixed with a few bluish-brown crystals will deposit. Bilirubin is insoluble in water and ether, slightly soluble in alcohol, but soluble in chloroform, turpentine, and benzol. On the addition of alkalies to a chloroformic solution of bilirubin it loses its orange-red colour. On passing a current of air through an alkaline solution of bilirubin, *biliverdin* is formed, which is deposited in green flocks on the addition of hydrochloric acid. *Bilifuscin* is another pigment, which can be obtained directly from the bile by evaporating a chloroform solution to dryness, dissolving the residue in alcohol, again evaporating, and shaking up the residue with ether and chloroform, separating the insoluble portion by filtration, and dissolving it in absolute alcohol, from which bilifuscin is precipitated in brown flocks on the addition of hydrochloric acid. These flocks are soluble in alcohol and alkalies, but insoluble in water, ether, and chloroform. According to Städeler, both biliverdin and

bilifuscin are formed from bilirubin by the assumption of water, thus :

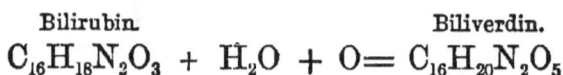

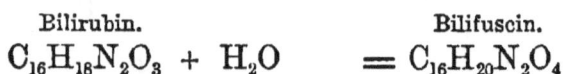

Bilirubin.   Biliverdin.

$$C_{16}H_{18}N_2O_3 + H_2O + O= C_{16}H_{20}N_2O_5$$

Bilirubin.   Bilifuscin.

$$C_{16}H_{18}N_2O_3 + H_2O   = C_{16}H_{20}N_2O_4$$

Although none of the bile pigments mentioned above
give any spectrum, there are other colouring matters
got from bile by treatment with stronger reagents than
those required for the separation of bilirubin, biliverdin, and bilifuscin.   Thus, by passing nitrous
vapours into an alcoholic solution of bilirubin a final
product of oxydation is obtained, which has been
named *choletelin.**   By pouring the alcoholic solution
into water after being thus treated, nearly all the
colouring matter separates in the form of flakes, which
dry up to a brown powder, which is soluble in alcohol,
ether, and chloroform.   It gives no play of colour with
nitric acid, but it yields a constant spectrum, which, in
an acid solution, gives one broad band, extending from
b to a little beyond F.   In alkaline solutions the band
is less refrangible.   Jaffé also isolated a pigment, which
gives the band at F, to which he gave the name of
urobilin.   R. Maly (*Ann. Chem. Pharm.*, clxi. 368,
clxiii. 77), by dissolving bilirubin in dilute soda or
potash ley, and adding sodium amalgam, the air being
excluded, no hydrogen was given off, but the dark
colour gradually lightened, and after two or three
days' action the solution acquired a yellow or bright
yellow colour, and then gave off hydrogen.   From
this liquid, hydrochloric acid separated a pigment,
which gave a spectrum identical with that of Jaffé's

* In the following I have adopted MacMunn's account of this
intricate and still obscure subject. "The Spectroscope in Medicine," p. 151.   Churchill.   1880.

urobilin.　Recent investigations seem to point to the conclusion that choletelin is the final oxydation product of bilirubin, and that urobilin is an intermediate stage ; and that while urobilin may appear in the urine, still, under normal circumstances, the pigment that appears in the urine is choletelin, though it may be absent in disease.　In addition to urobilin or choletelin, Stokvis (*N. Rep. Pharm.*, xxi. 123) has described another reducible product of the oxydation of bile pigment, which is formed as a secondary product, in most cases, of the oxydation of biliary colouring matter, whereby Gmelin's reaction is produced (§ 120).　It is colourless, or of a light yellow tint, soluble in water, alcohol, and dilute acids.　It differs from the bile colouring matter, and other oxydation products, in being insoluble in ether and chloroform, and not forming insoluble compounds with neutral or basic lead acetate.　When boiled with reducing agents in alkaline solutions this pigment yields a beautiful rose red, which gives in the spectrum a broad band in green.　In thick strata the band begins close to D and extends to *b*; in thin strata or dilute solutions it occupies only two-thirds of the distance between D and E, ending short of E.　This pigment does not exist in fresh bile, and it is not found in healthy urine.　MacMunn thinks that the appearance of the band of this pigment in the spectrum of the urine indicates grave disturbance of the system, as it appears only in those cases where there is undoubted disease of a severe character.　With regard to the connection between the colouring matters of the blood, bile, and urine, it is now generally held that the effete hæmoglobin is reduced in the liver to hæmatin and bilirubin, that a portion of the latter is passed out of the body with the fæces, but that another portion is converted into urobilin in the small intestine, and is reabsorbed, and by further oxydation is converted into choletelin.　The view that hæmatoidin and bilirubin

are identical is still open to question, though I think
the balance of evidence at present rather inclines to it.
To demonstrate the presence of bile pigment in any of
the secretions recourse is had to Gmelin's reaction. This
consists in slowly mixing (§ 120) a few drops of nitric
acid, containing traces of nitrous acid, with the suspected fluid, when a play of colours will be observed,
of which green is alone characteristic of the bile
pigment. If we examine this reaction by means of
the spectroscope, we find the solution gives a broad
shading in orange and yellow, and a broad band at F.
As the oxydation proceeds, the shading in orange and
yellow clears up, leaving only the band in F, the
spectrum of urobilin.

141. **Bile acids.**—The method for obtaining the
bile acids from urine has been described § 120. To
obtain them from bile we add excess of alcohol to the
alcoholic extract of bile, and filter. The precipitate is
dissolved in water, and the aqueous solution precipitated with neutral lead acetate. The precipitate is
removed, washed, dissolved in alcohol, and the lead
removed by precipitation with sulphydric acid. The
clear filtrate on standing will deposit crystals of·
glycocholic acid (§ 53). The mother liquor left after
the removal of glycocholic acid is precipitated with basic
lead acetate, and the precipitate treated in the same
way as directed for glycocholic acid, when taurocholic
acid will separate out as oily resinous drops (§ 54).
Both bile acids give a purple reaction (Pettenkoffer's
test) when mixed with strong sulphuric acid and
glucose. The best method of carrying out the test
is described § 120. The spectrum of Pettenkoffer's
test, according to MacMunn, gives a band outside D,
and broad band at E. Heynsius and Campbell, however, give the spectrum of sodium taurocholate with
sulphuric acid and sugar, as represented by three bands,
one between C and D, the next between D and E,

and the third near F (Pflüger's *Archiv. f. Phys.*, iv.,
497). The bile acids are furnished by the metabolism
of the albuminous constituents ; but whether directly
from the splitting up of the peptones, as Dr. Wickham
Legg considers probable, seeing the great dependence
of the bile-making functions on the glycogenetic function : or, as seems to me more likely, from the breaking
up of the effete albuminous matters of the body in the
liver, is doubtful. A portion of the bile acids (chiefly
the taurochloric acid) passes off by the bowels with the
fæces, partly unaltered, partly broken up into taurin,
cholic acid, and dyslysin. A portion, chiefly the glycocholic acid, is absorbed by the intestine. An experiment of Tappeiner (*Wiener Sitzgsber, Bd.* cxxvii.
1878, iii. abth) has shown that this absorption occurs
in the jejunum and ileum, and not in the duodenum.
Of the portion thus absorbed the ultimate fate is
unknown. A trace probably passes off with the urine
even in health, since Naunyn and Draggendorf have
proved the presence of bile acids in non-jaundiced
urine, whilst some portion of the taurocohlic acid is
probably oxydised, and furnishes the partially oxydised
sulphur product originally observed in the urine by
Ronalds, and which, in minute quantities, is always
present in normal urine (§ 114). Injection of the bile
acids into the blood produces very definite results. It
destroys the red blood-corpuscles ; if a few drops of
solution of bile acids are placed on a glass slide, and a
drop of blood be added, no trace of blood corpuscles
will be found on microscopic examination. Taurocholic
acid seems to possess this property in even a higher
degree than glycocholic acid. Bile acids, when injected,
cause rapid parenchymatous degeneration of the glands
and muscles. Their action on the heart is very marked,
causing slowing of the pulse. This phenomenon,
Dr. Wickham Legg (*Proc. Royal Soc.*, vol. xxiv., p.
442 ; 1876) thinks is not due to any influence through

the vagi, or direct action on the muscular walls, but
by their influence on the ganglia of the heart. Steiner
has confirmed Dr. Legg's view ; but he thinks that bile
acts upon only one of the cardiac ganglia, since he found
that by letting a drop of bile fall upon the back of a
frog's heart there was at once a cessation of the heart's
beat, whilst if it was placed on the fore surface of the
heart no change in the pulse took place for some time.
Injection of bile acids also produces a peculiar spasm
of the respiratory muscles, and the diaphragm remains
in a state of deep inspiration. They are said to lower
the bodily temperature. The urine after injection of
bile acids becomes dark-coloured, often with traces of
albumin, casts, and granules, but no blood corpuscles.
Frerichs considered the dark colour due to the conversion of the bile acids into bile pigment. Kühne
attributes it to the destruction of the blood corpuscles
by the bile acids acting on the hæmoglobin ; but Dr.
Wickham Legg has been unable to satisfy himself of
the presence of pigment, in many observations on
rabbits, and is inclined to believe the difference in
observation depends on the animal experimented on,
since dog's urine often gives the reaction (Gmelin's)
for bile pigment, even when the animals are supposed
to be in perfect health. Many physicians have been
led to regard the presence or absence of bile acids as a
diagnostic point, deciding whether the jaundice was
heptogenous or hæmatogenous in character. In the
former, it was said they were present, in the latter
absent. No such definite conclusion, however, is war-
ranted from the facts before us. Traces of bile acids
are to be found in normal urines ; the amount is
increased in all forms of jaundice, especially at their
onset. In cases due to obstruction, without great
destruction of liver tissue, they are often, especially at
first, considerably increased, gradually declining in
amount as the case goes on, till they often cease to

appear. In jaundice, with destruction of the liver tissue, as in acute yellow atrophy, hepatic abscess, rapidly growing cancer, they are present at first, but as the liver tissue becomes destroyed they disappear. Their presence or absence, therefore, does not distinguish between the nature of the jaundice, but only shows the stage which it has reached.

142. **Fats** which consist of cholesterin, saponifiable fats, and lecithin are to be separated and determined as directed for blood (§ 99). With regard to cholesterin, the chief fatty constituent of the bile, it is doubtful whether it should be regarded as formed in the liver, or is merely excreted from the blood. The latter view has been generally adopted, partly in consequence of Dr. Flint's views with regard to the supposed toxic influence this body has when its excretion is obstructed, giving rise to a condition he calls cholesteræmia. The reasons against accepting his views have been enumerated when considering the toxic conditions of the blood (§ 102). I am disposed to think that cholesterin is formed in the liver as well as in the brain, nervous system, etc., and that it is not an excretory product, properly so called, and that it fulfils a definite purpose in the organism ; that it is not excreted by the liver, but that some of the cholesterin formed in the organ is secreted with the bile. Cholesterin, mixed with bile pigment, is also the chief constituent of biliary calculi. For chemical reactions of cholesterin, *see* § 51.

143. **Salts.**—The inspissated bile extract is to be incinerated, and the estimation of the acids and bases made as directed for blood (§ 101). The chief base is soda, in combination with the bile acids, and which will in the incinerated ·residue be found as a carbonate. Sodium chloride is also abundant, and sodium phosphate ; then comes phosphate of lime and magnesia and chloride of potassium. Traces of iron

are always present, and are said to be increased when
this metal is taken as medicine. Minute traces of
silica, and also copper, are stated on good authority
to be constantly present.

144. **Biliary calculi** will be considered in
chapter vi., in the section referring to morbid concretions.

145. **Functional derangements of the
liver.**—In addition to the formation of bile, the
liver performs two other important functions, viz.,
(1) the formation of glycogen ; and (2) the metabolism of certain albuminous constituents of the
body. To these may be probably added another,
consisting of certain synthetical processes, by which
the carbohydrates are converted into fats, and the
peptones transformed into albumins prior to absorption into the blood by a process of dehydration
(§ 135).

(1) *Glycogenic function.—Diabetes.*—Although the
formation of glycogen is not restricted to the liver,
since it has been found in small quantities in other
tissues, still, undoubtedly, it is the chief seat of its
formation. The only exception to this statement is
during the first period of intra-uterine life, when the
liver is found free from this substance, whilst sugar is
found in the fluid of the allantois and liquor amnii.
During the latter period, however, of gestation the
liver of the fœtus contains glycogen, and sugar disappears from these fluids.* It has also been found
in the pulmonary tissues of hybernating animals.
In the consolidated lung of pneumonia, and in muscles
which have been kept long at rest, the quantity is
increased. With the resumption of activity, the
glycogen in the first case disappears, in the second

* In the liquor amnii of a diabetic patient, recently confined,
which Dr..John Williams brought to me for examination, no trace
of sugar could be found.

diminishes in quantity. Dr. Pavy has very ingeniously brought forward these facts to show that venous blood is favourable, and oxygenated blood is unfavourable, to its accumulation ; and, since there is no organ in the body supplied with venous blood in like manner to the liver, so, in correspondence, nowhere does glycogen exist to a like extent. In early fœtal life, however, the supply of venous blood to the liver from the chylopoietic viscera is quite insignificant compared with the oxygenated blood received from the umbilical vein, and so glycogen does not accumulate. As fœtal life advances this relationship becomes altered. In hybernating animals, and in muscles at rest, a reduced supply of arterial blood also must necessarily prevail. Dr. Pavy has, moreover, pointed out that in the consolidated lung, owing to its imperviousness to air, the venous blood of the pulmonary artery does not become oxygenated, but retains its venous character, and thus stands in the same position as the portal blood does to the liver. It is important to bear these facts in mind with regard to the origin of diabetes ; for, if a limited supply of oxygenated blood is favourable to the accumulation of glycogen, it is reasonable to suppose, if the opposite condition be present, and oxygenated or imperfectly de-arterialised blood be passed to the liver through the portal vein, rapid transformation of its amyloid substance into sugar will be accomplished, and glycosuria be the result. And this brings us to the question, what is the fate of the glycogen formed in the liver under normal conditions? The view originally suggested by Bernard, and generally accepted by the profession, was that the glycogen was immediately converted into sugar, and this, on reaching the general circulation, was destroyed, the destruction occurring chiefly in the peripheral capillaries and in the muscles, leading to the increased production of carbonic acid and water. Dr. Pavy, on the other hand, maintains that the liver is essentially a sugar-assimilating, instead of a sugar-forming, organ, and, when its assimilative action is properly exerted, so little sugar is allowed to pass into the general circulation that the quantity existing in arterial blood is insufficient to allow more than a mere trace to appear in normal urine. The glycogen is stored up in the liver cells, where it presumably undergoes a change which forms one of the links in the series leading up to the final issue; viz., the utilisation of sugar as a force-producing agent.

Now, according to Bernard's view, diabetes would depend on excessive formation of glycogen in the liver, by which an excessive amount of sugar, over and above its power of reduction into carbonic acid and water, would reach the circulation. Whilst, according to Dr. Pavy's view, diabetes is due to a failure of the assimilative function of the liver, which, instead of storing up glycogen, allows it to pass off as sugar, and, in proportion as it does so, the urine acquires a more or less marked saccharine character. Now, to return to the fact that excess of venous blood is favourable to the storing-up of glycogen, whilst oxygenated blood causes its disappearance. Dr. Pavy (*Proc. Royal Soc.*, June and Nov., 1875) has shown that, by introducing defibrinated arterial blood into the portal system, strongly-marked glycosuria was quickly introduced; and he also established the negative by employing defibrinated venous blood; that is to say, the glycosuria was due to the influence of the oxygenated blood, and not to any other part of the operation. This experiment shows that oxygenated blood reaching the liver through the portal vein is perversive of the proper action and instrumental in producing glycosuria. Dr. Pavy also induced glycosuria (11 grains of sugar to 1 oz. of urine) by artificial respiration ; and Tiffenbach has similarly succeeded in obtaining sugar in urine in curarised rabbits.

But in what manner in diabetes is the liver thus unduly charged with oxygen reaching the liver by the portal vein ? Dr. Pavy considers that the state into which the portal blood is thrown by vaso-motor paralysis affecting the vessels of the chylo-poietic viscera, is the key to the explanation of the saccharine condition of the urine in diabetes. He says, " It may be observed in the case of division of the sympathetic in the neck, that not only is there a hyperæmic condition of the ear, but that the veins contain much redder blood than natural. In fact, the blood passes with such velocity and in such volume through the affected part that it does not become de-arterialised. A similar state existing in connection with the vessels of the chylopoietic viscera will give what is sufficient to produce glycosuria. Without any new agent being brought into question, the simple passage of blood through the vessels in such a manner as to cause it to arrive in the portal vein in an im-perfectly de-arterialised condition will supply all that is wanted to account for the unnatural passage of sugar. In the vaso-motor paralysis, which, observation shows, is produced by lesions of the nervous system that give rise to glycosuria, we have a condition that leads to the presence of imperfectly de-arterialised blood in the portal vein, and in this condition of imper-fectly de-arterialised blood in the portal vein we have a circumstance that suffices to determine the escape of sugar from the liver in a manner to produce a diabetic state of the urine." Dr. Pavy also considers the bright red appearance of the tongue, so often noticed in severe cases of diabetes, is an evidence of this hyperæmic state ; the idea suggests itself from its appearance, that the blood is flowing through the

O

organ without being properly deprived of its arterial
character.   With regard to the nature of the nervous
lesion in diabetes that produces this vaso-motor
paralysis nothing definite has been determined.   Dr.
Dickinson has stated that he has found certain vascular and perivascular changes in the brain and medulla
oblongata of persons dying from diabetes.   His views,
however, have been challenged by pathologists, who
have found similar changes in brains of those dying
from other diseases besides diabetes, and that they
are not by any means a constant lesion in diabetes.
However this may be, I think the explanation of the
persistency of diabetes mellitus, a condition which
emphatically distinguishes it from glycosuria, will alone
be found to depend on some definite lesion of the
nerve centres, either of the cerebro-spinal or sympathetic system.   Indeed, it has been suggested that
as the vaso-motor nerves distributed in the sympathetic,
besides being connected with the spinal cord and
medulla oblongata, pass up to spots at the surface of
the brain, these possible vaso-motor centres should be
examined in all cases of diabetes mellitus.   That
question must be decided by the morbid anatomist; in
the meantime, the causes that induce glycosuria are of
greater interest to the pathological chemist.   In these
cases it is more than probable that the vaso-motor
paralysis is induced directly by the circulation through
the portal vessels of toxic agents.   Thus, Dr. Pavy has
shown that carbonic oxide, which gives to venous
blood a bright scarlet colour, and a similar spectrum
as oxygen does when introduced into the portal veins,
produces glycosuria in as marked a manner as arterial
blood ; and phosphoric acid, he has shown, has the
same effect.   George Harley has similarly shown that
chloroform, alcohol, ether, and ammonia produce
temporary glycosuria, so also does curare poison.
Although it has not been proved, it is far from unlikely

that uric acid may have the same effect, and thus
account for the frequent association of glycosuria with
the gouty state.   Even diminution of the alkaline
condition of the blood may play an important part in
producing this vaso-motor paralysis, since it has been
shown (Dr. Gaskell ; *Journ. Phys.*, No. 1, vol. iii. ;
1880) that whilst alkaline solutions cause powerful
contraction of the heart and capillaries, dilute acid
solutions have a contrary effect ; in this way one can
see how disturbance in the circulation between the
intestines and the liver, by permitting the acid chyme
to pass too rapidly into the portal vessels, might lower
the alkalescence of the blood in the portal circulation,
and thus cause temporary vaso-motor paralysis.   In
some anomalous forms of diabetes, disease of the pancreas has been noticed ; the cutting off of this powerful
alkaline secretion may have effect of diminishing the
alkalescence of the blood in the portal vessels.   Glycosuria is not unfrequently met with in those who have
had much mental labour, trouble, or anxiety.   Bernard
looked upon the appearance of sugar in these cases as
a salutary effort of nature to repair the injuries of the
organism, but it is far more probable that it depends
on a partial vaso-motor paralysis, an expression of the
general nervous exhaustion.

(2) *Abnormal disintegration.—Lithæmia.*—Formerly it was held that tissue changes depended on the
amount of oxygen taken in by the lungs, so that in
increased respiration a more intense combustion took
place, and metabolism was increased with the production of more carbonic acid and urea, whilst when
respiration was impeded oxydation was imperfectly
performed, and as a consequence many of the
intermediate products, as uric acid, oxalic acid, etc.,
were not burnt off, but were eliminated in an imperfectly oxydised condition.   Upon this view the late
Dr. Murchison founded his views regarding *lithæmia*.

" When oxydation," he says, " is imperfectly performed in the liver there is a production of insoluble
lithic acid (uric acid) and lithates (urates) instead of
urea, which is the soluble product resulting from the
last stage of oxydation of nitrogenous matter. Persons
who habitually enjoy the best of health are liable to
deposits of lithates (urates) in the urine after a surfeit
of food, or even after partaking moderately of one of
the fashionable dinners of the age. When more food
is taken into the blood than is necessary for the
nutrition of the tissues, the excess is thrown off by
the kidneys, lungs, and skin in the form of urea,
carbonic acid and water, or in the imperfectly oxydised
forms of uric acid and oxalic acid. Under these circumstances, an excess of work is thrown upon the
liver and the other glandular organs, and one
result is that a quantity of albumin, instead of being
converted into urea, is discharged by the kidneys in
the less oxydised form of uric acid or its salts. But
what in most persons is an occasional result of an
extraordinary cause, is in some almost a daily occurrence, either from the food being always excessive in
amount or unduly stimulating, or from some innate
defect of power, often hereditary, in the liver, in
virtue of which its healthy functions are liable to be
deranged by the most ordinary articles of diet."

The doctrine of lithæmia, or, as it ought more
properly to be termed, uricæmia, has found considerable favour with the profession, no doubt from the
powerful advocacy of such a distinguished clinical
teacher as Murchison; but though the numerous
symptoms detailed by him in connection with the
deposit of urates in urine are no doubt a matter of
common observation, his explanation of the causes
giving rise to the condition is not, I venture to think,
the correct one.

It is now known that uric acid is not a necessary

antecedent of urea, and that the latter body is
largely formed, independently of the liver, directly
from the kreatin of muscle and from leucin, the
result of pancreatic digestion, in the intestines. Nor
have we any evidence, except in the state of actual
gout, that, in human blood, uric acid has been ever
found; whilst the view is steadily gaining ground
that, with the exception of a small quantity
formed in some of the large glands of the body, and
in which it seems to be destroyed, uric acid is not
formed to any extent, and that the cyanogen residues
probably form urea directly without passing through
the intermediate stage of uric acid. Again, if uric acid
is found in excess in the urine because it has not been
oxydised into urea, we should certainly expect to find
the urea diminished in these cases; but this is not
so. In every analysis I have made myself, and in
others that I have referred to for the purpose of
ascertaining this point, I have found that, whenever
uric acid is increased, urea is likewise in excess, not
always proportionately, but still sufficiently decided as
to show there is no reduction in quantity.

The explanation I have to suggest as a cause for
the group of symptoms so admirably portrayed by
Murchison is, that they are due to a disturbance of
the nitrogenous equilibrium, brought about by an
increase of metabolic processes throughout the body,
owing to an intramolecular activity in the cells (§ 7).
This condition may be transient, and caused by in-
discretions of diet, nervous influences, or temporary
disturbances in some function of the organism; but,
when we have a persistent tendency to deposit urates,
accompanied by increased urea and phosphates, we
must look beyond mere defective oxydation or functional disturbance of the organ for an explanation,
and view it as a prelude of some grave constitutional
disturbance (often cancer, tubercle, or constitutional

syphilis), the taint of which produces increased molecular activity throughout the body. This question of lithæmia I have discussed from its clinical aspect in my work on " Morbid Urines associated with Derangements of Digestion," pp. 68—76, and to which I refer the reader, as that part of the subject cannot be introduced here.

As a converse to the conditions we have been speaking of is one characterised by a urine of low specific gravity, with no tendency to deposit uric acid or urates, and which is very deficient in urea, and associated with peculiar symptoms, which Sir Andrew Clarke has succinctly grouped together under the term " renal inadequacy." The term is an expressive one, and fixes in the mind the condition of the urine ; but I venture to think in this case we may go a step farther, and refer the inadequacy to failure in part of the metabolic processes going on throughout the body, and in part to defective assimilation of food by the digestive organs. Another evidence of abnormal disintegration is occasionally seen in cases of temporary albuminuria, and which are sometimes accompanied by the presence of peptones in urine. This condition is often of a transient nature, and dependent on marked disturbance of the hepatic function. In other cases, from being at first of a temporary and intermittent character, it becomes more and more persistent, as to give rise to a suspicion that it is a preliminary or early stage of granular kidney. Nor is it at all unlikely that this functional albuminuria may not at last terminate in organic disease. Hæmatinuria, too, is another condition which may be assumed to depend on functional derangement of the liver. In this disease (§ 120, page 161), only the colouring matter of the blood is present in the urine, and no blood corpuscles are to be found, whilst the colouring matter itself is apparently undergoing a change ; for, in addition to

the spectrum of oxy-hæmoglobin, there is also a third
band present, representing methæmoglobin. These
cases are usually attended with well-marked dyspeptic
symptoms and slight jaundice, and generally follow
exposure to cold or chill.   If, therefore, we regard the
liver as the seat of the disintegration of the effete
blood-corpuscles, with the formation of hæmatin, and
then of urobilin, it is not unreasonable to assume that
a sudden interference with the function of the liver
may temporarily arrest the process of the conversion
of hæmoglobin into bile and urinary pigment, and so
permit it to pass again into the circulation, to be
eliminated by the kidney.

146. **Pancreatic juice.**—A thoroughly reliable
analysis of the secretion of pancreas is wanting,
owing to the fact that the slightest irritation of the
gland causes considerable changes in the character of
the secretion.   Thus, Ewald, at the same period of
digestion in about equally large dogs, obtained sometimes a copious, sometimes a scanty, secretion, without being able to assign any cause for the difference.
It will, therefore, be sufficient to state that the
amount of solids in the pancreatic juice of dogs,
obtained from permanent fistulæ, ranges from 3 to 10
per cent.   Of these, the organic constitute about
two-thirds, and the inorganic one-third.   Of the organic solids, the most important are the ferments,
which act on starch, fat, and albumin; ordinary
albumin, an albumin precipitable by magnesium sulphate (casein or seroglobulin?); fatty matters; and
leucin and tyrosin.*   Of the inorganic constituents,
sodium carbonate is in relative excess of the other
salts.   The juice, when obtained free from the
products caused by the irritation of the duct, is a

---

* These bodies are not always present.   They seem to be
formed by the action of the proteid ferment on the albuminous
constituents of the secretion after it has been collected.

clear, somewhat viscid fluid, free from smell, having a marked alkaline reaction.

The ferments of the pancreatic juice can be obtained from the glycerin extract of the pancreas of a freshly-killed animal.

(1) *Action on starch* is most energetic, and surpasses that of the saliva (§ 131). The ferment has been isolated in a state of tolerable purity. It is stated that, after the pancreas has been exhausted by glycerin, if left exposed to the air for some time, it will again yield the diastatic ferment. From this, it is argued that the pancreas contains a body which by decomposition is converted into a ferment, analogous to the process that occurs in the liver by the conversion of glycogen into glucose (Liversidge; *Jour. Anat. and Phys.*, vol. viii., 1872).

(2) *Action on fats.*—The ferment that acts on fat has not yet been isolated. Its action, however, can be studied with the fresh juice, by mixing it with a little of a neutral fat in a test-tube, and placing the mixture in a water-bath (40° C.). At the commencement of the process the mixture is slightly alkaline, and neutral to litmus; but, as digestion proceeds, it becomes acid from the splitting up of the fat into glycerin and fatty acid.

(3) *Action on proteids.*—In dilute alkaline solutions, the ferment converts proteid bodies into peptones, similar to those formed by pepsin in acid solutions, dilute solution of sodium carbonate, 0·1 per cent., playing the same part as dilute hydrochloric acid does in gastric digestion. The points, however, in which pancreatic digestion differs from gastric digestion are, that the proteid elements do not swell up, become translucent, and fibrillate, as in the case of digestion with pepsin, but remain opaque, and apparently undergo conversion from the edges by a process of erosion. Again, the bye-product, anti-

albumose (or parapeptone) is of an alkaline character, resembling alkali, albumin, or casein, and not acid albumin or syntonin, as in gastric digestion.   Lastly, the continued action of pancreatic ferment leads to the formation of leucin and tyrosin from the breaking up of the hemipeptone.   The following table shows in a graphic form the changes effected on proteid substances respectively by gastric and pancreatic digestion :

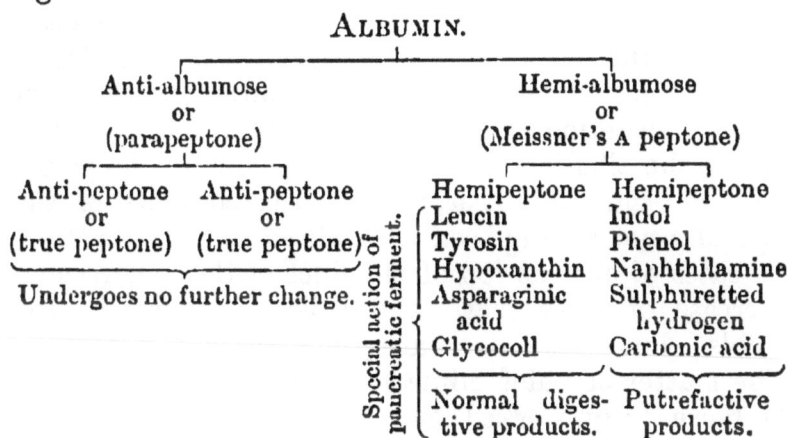

ALBUMIN.

Anti-albumose or (parapeptone)		Hemi-albumose or (Meissner's A peptone)	
Anti-peptone or (true peptone)	Anti-peptone or (true peptone)	Hemipeptone Leucin Tyrosin Hypoxanthin Asparaginic acid Glycocoll	Hemipeptone Indol Phenol Naphthilamine Sulphuretted hydrogen Carbonic acid
Undergoes no further change.		Normal digestive products.	Putrefactive products.

Special action of pancreatic ferment.

Thus the final action of the pancreatic ferment is to break up the hemipeptone formed by its own digestive action, and the hemipeptone formed by gastric digestion, into a series of bodies related to the group of fatty acids (leucin = amido-caproic acid) and the aromatic series (tyrosin = amido-propionic acid in which one atom of H is replaced by oxy-phenol).   The formation of indol, phenol, etc., is, however, regarded as due to putrefactive changes, and not to normal pancreatic digestion, since the production of these substances is simultaneous with the appearance of organisms in the solutions.   These organisms are probably taken in with the food and find a habitat in the intestines ; they do not exist in the pancreas.   The chemical reactions of the true peptone formed by pancreatic digestion are similar to that formed by the gastric juice (§ 135, page 186). The pancreatic ferment dissolves mucin; and, according to Nencki, converts gelatin into a peptone and glycocoll. It has, like pepsin, no action on lardacein, nuclein, or keratin.

The pancreas is an organ about which the pathological chemist has little to say as yet. Fatty degeneration and atrophy of the gland has been noticed by Bright, Frerichs, Catani, and numerous other observers, in some cases of diabetes. By some it is attributed as the result of that disease. At page 211 I have suggested how possibly diminution of the pancreatic secretion may induce glycosuria, by diminishing the alkaline reaction of the blood in the portal vessels from the absorption of the contents of the jejunum not having been neutralised by this secretion. In cases of obstruction of the duct of the pancreas an increase of fatty matters has been noticed in the stools; but there are exceptions to this statement. The matter of chief clinical interest, which ought to be more fully worked out, is the formation of urea from the leucin derived from pancreatic digestion, and the formation of indol, and its appearance as indican in the urine. With regard to the latter, it has been shown that as salicylic acid puts a stop to the formation of indol in albuminoid solutions of pancreatic juice, so the internal administration of the same acid reduces the amount of indican in the urine, which decreases exactly in proportion as the quantity of phenol increases. Indican is also met with in the urine after ligature of the small intestines, and in obstructions and other affections of the intestines in disease. The indigo sometimes deposited in a free state in urine, sweat, and urinary calculi, is undoubtedly derived from the indol formed in the intestines by pancreatic digestion. The indol apparently is converted into indican in the alkaline blood, which

is converted into indigo, if present in excess,
when it comes in contact with highly acid sweat or
urine.

147. **Intestinal digestion.** — Very little is
known regarding the composition of the secretion
furnished by the intestinal glands. According to
Thirry the *succus entericus* is a yellowish opalescent
fluid of alkaline reaction, having a specific gravity of
1·011, and contains about 2·5 per cent. of solids.
It dissolves fibrin, but it is a question whether it has
any action on other proteid bodies. Its diastatic
action on starch is doubted. It is said, however,
to convert cane sugar into grape sugar and invert sugar, and to set up lactic acid fermentation.
Absorption of the digested products takes place
throughout the intestinal canal, but the process
must be considered with reference to particular
tracts :— (1) The duodenum ; (2) the jejunum
and upper part of the ileum ; (3) the lower part
of the ileum and the large intestine. (1) *In the duodenum.*—In the stomach most of the diffusible sugars,
and some of the peptones, pass directly into the
gastric veins ; but the remainder, consisting of parapeptone, the undigested albumin, the starchy principles which have as yet escaped conversion, and the
oleagineous matters, pass through the pyloric orifice
into the duodenum as acid chyme. On reaching the
biliary orifice a rush of bile takes place, which causes a
precipitation of the parapeptone, at the same time the
pancreatic secretion is poured forth in abundance.
The action of these two alkaline secretions has the
effect of first neutralising the acid and then rendering
the contents of the intestine alkaline. In the meantime, the emulsionising and saponification of the fatty
matter is proceeding, and the chyme passes into (2)
the jejunum and upper part of the ileum as chyle. It
is here that the final digestion of the food is effected,

the pancreatic juice converting the undigested albumin and some of the gastric peptone into anti-peptone and hemipeptone, which last is converted into leucin, tyrosin, etc. The starch is converted into grape sugar, so is, probably, part of any cane sugar present, whilst a portion commences to undergo lactic fermentation. The fats meanwhile are thoroughly emulsified and saponified. These products are quickly absorbed, the peptones, the glucose, and a dextrin-like body, passing chiefly in the direction of the portal vessels, whilst the emulsioned fat is taken up by the lacteals. (It is a question whether the saponified matters are taken up, or whether saponification is only subsidiary to emulsion.) In the absorption of fatty matter the bile acids undoubtedly play an important part by aiding their passage through the absorbents. As the fluid contents pass onward they gradually lose their nutritious constituents, till at (3) *the lower part of the ileum and large intestine* they become more consistent and contain little besides the insoluble residue of the food and the putrefactive products, indol, etc., of pancreatic digestion. The reaction, too, here again becomes acid from the lactic acid fermentation set up in the intestine. Although the process of absorption is very inconsiderable, still it exists to a very appreciable extent, as is decidedly proved by the beneficial result nutritive enema have for a time on patients who cannot otherwise be fed.

148. The **fæces** consist of that portion of the food which is not taken up by the absorbents, and is discharged from the body mixed with some of the products of the biliary and intestinal secretions. Their amount must necessarily depend on the nature of the food taken, and on the energy of the digestive powers. The quantity passed by a healthy adult may be, however, stated at from 7 to 9 ounces daily.

The colour ought to be a rich brown, and the surface
of the motion moist and slightly slimy. The odour
varies considerably in different persons ; in some, even
in the healthy, it is particularly strong and offensive.
This odour is chiefly derived from the putrefactive
changes occurring in the intestine, and partly to a
secretion of the glands of the large intestine. Those
who make it their duty to inspect the stools of their
patients will soon learn to recognise an odour *sui generis*
attached to many disorders ; thus, it is easy to distinguish by smell alone a dysenteric from a typhoid
stool, etc. The following is an approximate analysis
of the fæces of a healthy adult : Water, 77·3 ; Solids,
22·7 ; mucin 2·3, proteids 5·4, extractives 1·8, fats
1·5, salts 1·8, resinous, biliary, and colouring matters
5·2, insoluble residue of food 4·7.

(1) *Albumin.*—A small quantity of coagulable
albumin is always present in normal fæces, whilst in
cases of dysentery, typhus, and cholera, the quantity
passed is considerably increased.

(2) *Extractives.*—Besides the occasional presence
of leucin, certain fatty acids, substances called excretin, and stercorin, are present. The following is an
account of these substances and mode of separation,
though many are sceptical as regards their existence
in a definite form, but consider them to be modifications of cholesterin. (a) *Stercorin.* First described
by Dr. Flint, who assumes that under ordinary
circumstances about 0·6 gramme is excreted daily.
Dr. Flint directs the fæces to be evaporated to
dryness, pulverised, and exhausted with ether.
The etherial solution is then passed through animal
charcoal, fresh ether being added, until the original
quantity of ether extract has passed through. The
filtered etherial solution is then evaporated, and
the residue treated with boiling alcohol. The
alcoholic solution is evaporated, and the residue

treated with a warm solution of caustic potash to dissolve out all the saponifiable fats. The mixture is then diluted with water, thrown on a filter, and washed till the droppings are clear and neutral. The filter is dried, and the residue washed out with ether. The etherial solution is then evaporated, and the residue treated with boiling alcohol, the residue of this solution yielding stercorin. Stercorin, when thus obtained, appears as a clear, amber-coloured, oily substance, in which thin, needle-shaped crystals, frequently arranged in bundles, and having their borders split longitudinally, appear in the course of a few days. Stercorin is neutral, soluble in ether and hot alcohol, insoluble in water and solutions of potash ; it is distinguished from cholesterin by having a lower melting point, viz. 38° C. Treated with strong sulphuric acid it gives a red colour. Dr. Flint considers that stercorin is formed by a modification of cholesterin in its passage along the intestinal canal ; since a comparison of the total quantity of cholesterin contained in bile with the quantity of stercorin actually discharged shows a correspondence.

(*b*) *Excretin.* This principle was obtained by Dr. Marcet, together with excretolic acid, from fæcal matter. The fæces are first dried and exhausted with boiling alcohol, and the alcoholic solution concentrated, filtered, and allowed to stand ; after some time a granular, olive-coloured, fatty acid, excretolic acid, is deposited. This substance melts at 25°, is insoluble in water and in solutions of potash, and in cold alcohol ; its composition has not yet been determined. The excretolic acid must be removed by filtration, and the filtrate treated with milk of lime, which throws down a brown precipitate ; this is dried and exhausted with ether, which yields crystals of excretin. The crystals form delicate, silky, four-sided prisms, insoluble in water and solutions of potash, very soluble in ether; they melt at 95°, and have an alkaline reaction.

3. The *fats* contain saponifiable fats, and a very considerable proportion of cholesterin.

4. The *inorganic constituents* contain a very considerable amount of magnesium and calcium phosphate, the former chiefly in the form of triple phosphate. The fæces are the only instance in the animal tissues and fluids in which the magnesium salts are relatively in excess of the calcium salts; this is owing to more lime than magnesia being absorbed in the intestines. The potassium salts are also relatively in excess of the sodium salts. The biliary acids appear in the fæces in an altered condition, as dyslysin, cholalic, and cholodinic acids. The insoluble residue consists of undigested muscular fibre, the outer envelope of vegetable cells and fibres, partially dissolved starch, cells of cartilage, and fibres of elastic tissue, etc. Fæces sometimes contain a ferment like pepsin, and one that has a diastatic action on starch; both are probably derived from the digestive secretions. Meconium, or the fæcal matter at birth, consists almost entirely of biliary matter and mucus. Thus, a sample of dry meconium yielded : Biliary matter, 15·6 ; cholesterin, 15·4 ; mucus epithelium and salts, 69 parts in 100.

We have very few reliable analyses of fæces in disease, few having courage to pursue diligently such an unpleasant business. As a rule we learn much by regular inspection; and if the plan adopted at the London Hospital, of placing them in large deep conical vessels, and covering the mouth of the vessel with a thick glass plate, inspection can be made without unpleasantness being occasioned, especially if with loose stools a little ether is floated over them. In liquid stools the more solid contents gravitate first, and so on, so that a rough analysis can be

made. If a more intimate acquaintance with the
composition of the fæces is desired, they must be
submitted to regular analysis; viz. : (1) By ascertaining the amount of water and solids (§ 92);
(2) by agitating in water for some time a definite
weight of fæces, filtering the aqueous solution, and
coagulating the dissolved albumin, and weighing
)ι as directed (§ 118, page 145); (3) extracting with
ether to remove fatty matters and cholesterin (§ 99);
(4) after the removal of the cholesterin, extracting
successively with boiling alcohol and chloroform to
remove biliary matters (page 198); (5) whilst the salts
are estimated after incineration, as directed (§ 101).

149. **Intestinal concretions** will be considered in chapter vi., in the section referring to
morbid concretions and calculi.

150. **Gases in stomach and intestines.**—
As has been well observed, though yeast fungi are
continually being taken with the food, as in bad beer
or bread, and are thus brought in contact with the
saccharine and albuminous matters of the food, which
are capable of fermenting in the stomach, fermentation does not occur unless another condition is added.
The ferment must have time and opportunity for
developing itself. Under ordinary circumstances it is
so rapidly removed from the stomach, together with
fermentable material, that the process has no time to
commence. The conditions, therefore, that favour the
development of fermentation are those which retard
digestion either by mechanically obstructing the
onward passage of the food, or from an abnormal
condition of the digestive secretions, or the indigestible
nature of the food itself. From experiments, we learn
that under normal circumstances the gases found in
the *stomach* consist of oxygen, nitrogen, and carbonic
acid, but no hydrogen, which we would expect to find
if the gases of the stomach in health were formed by

lactic acid fermentation. It is probable, therefore, that of the gases obtained from the stomach under normal conditions, the first two are derived from the air swallowed with the food, whilst the latter is derived by diffusion from the blood. In the *small intestine*, however, acetic and lactic acid fermentation commences, as is shown by the preponderance of carbonic acid gas, and the presence of hydrogen. The following table gives the result of Planer's analysis of the gases of the stomach and intestine respectively:

Gas.	Stomach.		Small intestine.	
	Meat.	Bread.	Meat.	Vegetable diet.
CO_2	25·20	32·91	40·1	47·34
N	68·68	66·30	45·5	3·97
O	6·12	·79	Trace	—
H	Nil	Nil	13·86	48·69

The steps that occur in this process of fermentation are shown in the following table, thus :

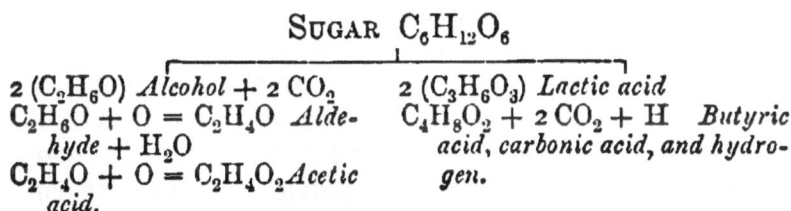

$$\text{SUGAR } C_6H_{12}O_6$$

2 (C_2H_6O) *Alcohol* + 2 CO_2 2 $(C_3H_6O_3)$ *Lactic acid*
$C_2H_6O + O = C_2H_4O$ *Alde-* $C_4H_8O_2 + 2 CO_2 + H$ *Butyric*
   hyde + H_2O *acid, carbonic acid, and hydro-*
$C_2H_4O + O = C_2H_4O_2$ *Acetic*   *gen.*
   acid.

The occurrence of this lactic and butyric acid fermentation in the small intestine in health, suggests a way in which the carbohydrate constituents of food may become converted into fat; for, by this lactic and butyric acid fermentation, the sugar is converted into members of the fatty acid series. The extent, however, to which this fermentation is carried on in

P

health is probably small, since if it occurred largely in
the intestine we should have a considerable quantity of
free hydrogen excreted by the lungs or bowels, which is
not the case. The fermentative changes reach their
highest point in the large intestine, so much so as to
render its contents acid, in spite of the alkaline
character of the secretion from its walls. Here, in
addition to hydrogen, we have a considerable quantity
of marsh gas (CH_4) developed with sulphuretted
hydrogen (H_2S), from the decomposition of the
albuminous and other sulphur-yielding elements of
the food.

In disease, however, excessive fermentative changes
of the food may occur, leading to the production of
enormous quantities of gas and the formation of
various intermediate products, as we have seen, such
as alcohol, aldehyde, and acetic acid on the one hand,
and of lactic and butyric acid on the other. Sometimes it is a large quantity of gas that is formed, at
another time an excess of acid. Thus, Ewald speaks
of a patient who pithily observed that " there was
sometimes a vinegar factory and sometimes a gas-
works in his inside ; " in fact, at one time alcoholic
fermentation led to the formation of acetic acid, at
another, the butyric acid fermentation produced
hydrogen and carbonic acid. It is often difficult to
distinguish clinically between the different forms of
flatulent distension which arise ; but we receive considerable aid if we are careful to discriminate between
those forms where flatulence is the only symptom and
those where it is associated with acidity, and also by
taking into consideration the period with regard to
digestion at which these symptoms develop. Thus,
there are some persons, chiefly females, who, immediately on taking food, complain of flatulent
distension without acidity ; the wind they bring
up is inodorous. In these cases, the gas does not

apparently result from fermentative changes, but is
probably derived by diffusion from the blood under
nervous influences. When the flatulency is accompanied by a slight degree of acidity, and sets in about
an hour after food, and the risings are simply acid,
and the eructations comparatively inodorous, acetic
and carbonic acid fermentation of the amylaceous
and saccharine materials of the food is indicated.
When the risings are distinctly rancid, it is evidence
that lactic acid fermentation of the nitrogenous principles is progressing. This form of fermentation is
usually the most obstinate and severe, since it may
continue independently of food by the decomposition
of the mucus in the stomach and the intestinal canal,
so that flatulence may persist even when the stomach
is kept empty.

Flatulent distension of the stomach and intestines
often arises in nervous states of the system, apparently, however, quite independently of any fermentative changes occurring in the alimentary canal.
Indeed, it is quite impossible to account for the
enormous quantity of gas, which consists largely of
carbonic acid, often discharged through the mouth on
a perfectly empty stomach by hysterical and hypo-
chondriacal patients, except on the supposition that it
is diffused from the blood.

The disturbances caused by fermentative changes
in the stomach are not limited to that organ. The
acid products formed in it, together with the undigested residue of the food, pass on into the intestines, and excite more or less pain and diarrhœa.
Again, fermentative changes may occur chiefly in the
intestines, and only a slight degree in the stomach.
In this case, which is associated with a greater or less
degree of chronic intestinal catarrh, a constipated
condition of the bowels generally exists; for, though
there may be frequent loose, slimy, and offensive

discharges from the bowels, yet a purge never fails to bring away accumulated masses of fæcal matter. The whole of the intestinal tract may be affected, or only part of it. Some writers have asserted that the evil effects of fermentative changes are more felt in the small intestines than in the large, and that catarrh of the small intestines is generally associated with oxaluria. The flatulence may distend the whole intestinal tract, but one part of it is generally distended more than another; and circumscribed swellings occur, chiefly in the right and left hypochondriac regions, causing pain over the region of the liver and stomach, spleen, or kidneys, and leading the patient to suspect disease of these organs. But the mischief resulting from excessive formation of acid in the stomach and bowels is not limited to mere disturbance of digestion, injurious effects making themselves manifest on the system and on the general nutrition of the body when the morbid condition has been present for some time.

Thus, Beneke* has pointed out that the increased production of lactic and butyric acids in the alimentary canal is frequently associated with oxaluria (§ 112, page 130), since, as he thinks, the excessive formation of these acids prevents the development of the red corpuscles, so that oxydation is insufficiently performed. A catarrhal condition of the mucous membrane of the intestines he also pointed out as being frequently found accompanying this condition ; he does not, however, consider it as being a proximate, but only a determining, cause of the disorder. Whilst endorsing Beneke's statement that deposits of oxalate of lime are met with in persons suffering from dyspepsia, attended with excessive formation of lactic and butyric

* "Zur Phys. und Path. des Phosphors und Oxalsaure Kalkes." 1850. "Zur Entwicklungsgeschichte der Oxalurie." 1852

acids, I do not consider his explanation to be the correct one, since, in these cases, I believe a catarrhal condition of the mucous membrane of the digestive canal to be the proximate cause, which, by hindering the onward passage of the food, favours fermentative changes and the production of lactic and butyric acids.

Again, the highly acid fluid containing the imperfectly digested products of gastric digestion passing into the duodenum excites more or less catarrh of that portion of the intestine, and the discharge of bile is interfered with; hence persons suffering with flatulent dyspepsia have usually sallow complexions, complain of pain in the hepatic region, and suffer frequently from so-called "bilious attacks." The absorption of the vitiated products of digestion, together with some of the free acid, produce many general and remote disorders of nutrition, so that a condition of debility and exhaustion is speedily induced.

151. **Detection of arsenic, antimony, etc., in the viscera.** — A plan of procedure for the examination of vomited matters by which the student or practitioner can ascertain the nature of the poison, without the expenditure of much time or the employment of elaborate apparatus, has been detailed § 136, page 191. Of course, it is understood that these investigations are only preliminary to a more searching investigation in the chemical laboratory at the hands of an expert. In addition to the examination of the vomited matter ejected during life, we should also make an examination of the tissues, especially of the viscera, after death, especially if, as is sometimes the case, the vomited matters have been thrown away without having been examined, and the patient having lived some hours after the poison has been absorbed from the stomach into the tissues. For

this purpose portions of the liver and stomach must be divided as finely as possible and placed in a porcelain dish, and a mixture, about twice the quantity of the organic matter employed, consisting of six parts of distilled water to one of hydrochloric acid, is added, and the whole warmed for about an hour. After this, small fragments of potassium chlorate are to be dropped into the mixture from time to time, and the mixture kept constantly stirred, till the solid matter has almost completely disappeared. The mixture is then filtered through fine linen, the insoluble matter left on it being kept for further examination, and the acid filtrate divided into three parts : (1) Place in the acid mixture a strip of perfectly pure copper and boil for twenty minutes ; if there is a deposit on the copper examine for *arsenic, antimony,* or *mercury.* Remove the strip of copper, wash it with a little distilled water to which a few drops of ammonia are added, dry it between folds of blotting paper, then when quite dry place it at the bottom of a narrow glass tube (German glass), and apply heat to the lower portion of the tube, taking care that the upper end remains cool, and placing the finger lightly over the mouth of the tube, so as to keep the volatilised matters within it. If *arsenic* forms the crust on the copper, then arsenious acid will sublime and be deposited at the upper end of the tube, and this deposit under a low power of the microscope will be found to consist of sparkling octohedral crystals. Break off the portion containing the deposit and boil it in a test-tube for some minutes with distilled water. Test aqueous solution with (*a*) few drops of silver ammonium nitrate, which gives a bright yellow precipitate, soluble in ammonia and nitric acid ; (*b*) solution of cupric ammonio-sulphate, which gives a pale apple-green precipitate. If arsenic has been found it will be as well to take a fresh portion of acidulated filtrate and submit it to Marsh's test. If the crust

deposited on the copper is caused by *antimony*, it does
not, by heating, yield a crystalline, but an amorphous
deposit, and this, on the application of a greater
degree of heat than was required to volatilise the
arsenic.   To prove the presence of antimony, break
off the upper end of glass tube, boil it with distilled water very slightly acidulated with hydrochloric
acid.   Through this acid solution pass a stream of
sulphydric acid, when an orange-coloured precipitate
will fall.   If the deposit on the copper is caused by
mercury, on volatilisation distinct globules will form
in the upper part of the tube; for corroboration apply
tests given § 132, page 177.   If no result has been obtained by boiling the copper in the acid filtrate, take
a fresh portion (2) of the acid solution and warm it
and pass a stream of sulphydric acid through it.   A
black precipitate indicates *copper* or *lead*.   If copper,
on dipping the blade of a knife in the acid solution,
which should be concentrated, it will be deposited
on the surface.   Ammonia added gives, with solution
of copper, a light blue precipitate, soluble in excess
of ammonia with the formation of deep blue solution.
Potassium ferrocyanide gives, with solution of cupric
salts, a reddish-brown precipitate.   This test is
employed to show that all the copper has been deposited from the standard Fehling solution in quantitative estimation of glucose (§ 119, page 153).   If the
black precipitate is not caused by a salt of copper, then
test solution for lead as directed § 129, page 172.
If no precipitate is caused by sulphydric acid a
solution of ammonium hydro-sulphate is to be added;
if a zinc salt is present a white or yellowish-white
precipitate will be thrown down.   (3) The acid
filtrate is to be concentrated to a small bulk, and is
then rendered alkaline by potash and shaken with five
times its volume of ether.   The etherial solution is
then removed and allowed to evaporate spontaneously

in a small glass dish and the solid residue examined
for strychnine or morphia by dissolving it in a little
dilute hydrochloric acid, rendering this solution alkaline by sodium carbonate and allowing the mixture
to stand till a crystalline precipitate forms. This
is to be collected on a small filter and washed till
the washings are no longer alkaline. Then divide the
filter into two parts and lay them on white porcelain
plates and test respectively for morphia (§ 70) and
strychnia (§ 71). The filtrate and the washings are
mixed together and evaporated to dryness in a water-
bath, and the dry residue warmed with alcohol ; the
alcoholic solution is then to be evaporated to dryness
and the tests for morphia and strychnia applied.

152. **Estimation of nitrogen.**—In some physiological and pathological enquiries, we wish to compare
the quantity of nitrogen passed out of the system with
the fæces and urine, with the amount of nitrogen taken
in with the food. For these investigations the soda-
lime process, devised by Voit, is the best. For this
purpose a small tubular retort, the bulb of which is
about 6 centimetres deep, and 3·5 centimetres broad,
with the neck bent at right angles about 10 centimetres from the bulb, and drawn out into a fine tube
8 centimetres long, and 0·3 centimetres in diameter,
has placed in it some soda-lime, recently heated to
redness, to the depth of 1·5 centimetres. The narrow
tube of the retort is then fitted to a glass flask (of
150 cc. capacity) by means of a perforated cork. The
delivery tube of the retort should pass down almost
to the very bottom of the flask. By means of a
second hole in the cork a glass tube is fixed which is
adjusted so as to be above the level of the fluid. The
apparatus being arranged, 100 cc. of standardised
dilute sulphuric acid is poured by means of the glass
tube into the flask, and then 5 cc. of urine (or in the
case of fæces or food 5 grms., weighed dry, and

afterwards mixed in 5 cc. of water) are poured on the
soda-lime. The mixture at once becomes warm and
ammonia is disengaged, which passes over into the
flask where it bubbles up through the dilute sulphuric
acid. At this point of the proceeding there is a
tendency for the sulphuric acid to be sucked up the
tube into the retort. To prevent this the retort is
held over a spirit lamp and heat very cautiously
applied, care being taken that while the heat is
sufficient to prevent the sulphuric acid rising in the
tube it does not cause too violent disengagement of
gas. When the whole of the water has passed over
from the retort into the flask, and the gas is passing
off regularly and steadily, the temperature must be
raised. This is done by placing wire gauze round
the bottom of the retort and applying strong heat,
till the mixture in the retort becomes nearly white.
When this is so, bubbles of gas are no longer disengaged, and the sulphuric acid begins to rise in
the tube. The heat must now be withdrawn and
the glass stopper taken out of the retort. The
flask is detached and the contents poured into a
beaker, a few drops of litmus solution (page 70) added,
and the non-saturated sulphuric acid measured with an
equivalent quantity of a standard soda solution. Thus,
of 100 cc. of the sulphuric acid solution 20 cc. are found
to be saturated, as ammonia sulphate, and as 1 cc. of
sulphuric acid = 0·00425 grm. of ammonia, or
0·0035 grm. of nitrogen; then 5 cc. of urine contain
·0035 × 20 = ·0700 grm. of nitrogen. Consequently
if 1200 cc. of urine have been passed in the 24 hours

$$\frac{·0700 × 1200}{5} = 16·8 \text{ grms. of nitrogen (258 grains).}$$

And so with fæces, if 80 grms. represent the amount
of dried fæces passed in the 24 hours, of which 5
grms. are distilled with soda-lime, and the resulting
ammonia saturates 25 cc. of the sulphuric acid

solution, then ·0035 × 25 = ·0875 grm. of nitrogen
for every 5 grms. of dry fæces, and consequently
$\dfrac{0875 \times 80}{5} = 1\cdot6$ grms. of nitrogen (24·6 grains) passed
off by the bowels in the twenty-four hours. It was by
this method that the late Professor Parkes showed the
close parallelism that existed between the entrance
and exit of nitrogen. Thus, in one case, with an
entrance of 270 grains of nitrogen, 254·04 were passed
off by the kidneys, and 27·74 by the bowels, making a
total exit of 281·78 grains. In a second case the
entrance amounted to 302·6 grains, and the exit
312·8 grains, of which 296·2′ grains passed off with
the urine, and 16·6 by the bowels. Ranke also found
that with an entry of 296 grains of nitrogen, 26·23
grains were passed by the bowels, and 281 grains by
the kidneys, making a total of 307·23 grains of
nitrogen.

The standard solution of sulphuric acid required
for the process is made as follows :—12·6 grammes of
oil of vitriol are weighed and diluted up to a litre of
distilled water ; the quantity of sulphuric acid in each
20 cc. should be determined by barium chloride
solution, so that each 100 cc. of the dilute acid should
contain 1·0 gramme of sulphuric acid, and this
corresponds to 0·425 gramme of ammonia, or 0·35 ,
gramme of nitrogen. Consequently 1 cc. contains
0·00425 gramme, or ·0035 gramme of nitrogen.

The sodium hydrate solution must be standardised
so that 20 cc. of it should exactly neutralise 20 cc. of
the dilute sulphuric acid solution.

CHAPTER VI.

MORBID PRODUCTS.

152. Urinary and renal calculi.—With regard to the origin and mode of formation of urinary calculi, it may be stated that the nuclei, except in the case of foreign bodies introduced from without, are formed in the kidney, where they may be retained to form a renal calculus, or pass down the ureter into the bladder, and so become vesical. Their increase in size depends on additions made to the original nucleus by the constituents of the urine, and these additions may be either of the same nature as the nucleus itself, or may be formed of other urinary constituents, which may become subsequently in excess, or may be deposited by alterations of the reaction of the urine. In considering, then, the history of any given stone, we have first to ascertain the nature of the causes that led to the formation of the nucleus, and then trace the growth of the stone as exhibited by the chemical composition of its successive layers. Various views have been advanced to explain the formation of calculi. The ancient authors held that stone was formed in the urinary organs by a kind of slime baked by the heat and dryness of the parts, just as a portion of soft clay may by external heat be turned into brick or tile.* This view maintained until Marianus Sanctus† pointed out that calculi were of two kinds, the one no doubt formed by heat, but others by cold and humidity as marble was formed. Paracelsus was the first who

---

* Hippocrates (περι αἐρων, ὑδατων τοπων, cap. ix.).
† De lapide renum et vesicæ, in thesauro chirurgiæ, Petri Uffenbachii. Folio. p. 903. Francof. 1610.

submitted calculous matter to chemical analysis (1530). He demonstrated that it was composed of organic matter or nutritive principle, an earthy principle, and a volatile salt ; and he thought calculous matter was of the nature of *tartar*, caused by the union of the nutritive principle and the saline spirit which coagulated the earthy matter.   This view, with some modifications, was held by Van Helmont, and the iatro chemists who followed, and it ultimately resulted in the doctrine of a " concreting acid," so that when Scheele in 1776 discovered uric acid, it was regarded as being the formative principle of stone, and was in consequence designated lithic acid.   This view was proved erroneous by Wollaston's discovery of calcium oxalate, ammonio-magnesium phosphate, and calcium phosphate, as constituents of some stones.   These discoveries, however, gave too great a prominence to the chemical origin of stone, and the idea gained ground that urinary calculi were the result of a peculiar diathesis wherein uric acid, oxalic acid, or the phosphates were formed in excess in the system, and were eliminated in such quantities that they were precipitated in the urinary passages.   The first correction of this view was made by Prout with regard to uric acid, and Bence Jones with respect to phosphates, both of whom showed that these substances for their precipitation need not be in excess, but that they were precipitated even when present in their normal quantities by changes taking place in the reaction of the urine. About this time Rainey published his researches on molecular coalescence (§ 11, page 26), which drew attention to the important part played by the mucus of the urinary passages in furnishing the colloid medium in which the precipitated matters were moulded.   In fact it was the slime of Hippocrates, the " gros humeurs gluans, espais et visqueux " of Ambrose Paré (1564).   In all researches, therefore,

into the origin of stone these two branches of enquiry
must be taken in hand together, viz., the nature of
the precipitated matters and the supply of colloid
material. With regard to the former our knowledge
is now pretty definite, and the causes of their deposition will be found related § 111, page 125, § 113,
pages 133 and 134. Thus we have seen that uric
acid and urates are deposited from highly acid urine ;
calcium phosphate from urine, alkaline from the fixed
alkalies ; ammonio-magnesium phosphate (triple phosphate) from urine, alkaline from volatile alkali derived
from the decomposition of urea ; and calcium oxalate
probably from an acid fermentation of mucus in the
urinary passages as well as from the calcium oxalate
derived from the system. With regard to the nature
of the colloid medium various views have been expressed. Thus, Dr. Owen Rees is in favour of a gouty
catarrh. Among the many evils arising from the
gouty diathesis is a tendency to this kind of action on
the part of mucous surfaces. Others have suggested
some chronic sub-inflammatory condition. Among the
Germans a specific catarrh (*stein-bildenden catarrh*)
has found favour. But if stone originated from
catarrhal or inflammatory conditions it would be more
frequent than it is, since catarrhal conditions associated with deposits of urinary constituents often subsist together without stone resulting ; for instance, in
nephritis, where, with abundance of tube-casts, blood
and epithelium poured into the urinary passages, we have
frequently co-mingled urates and uric acid, deposited
from the concentrated urine, yet stone is not recognised
as a clinical sequence of acute nephritis. Again,
calculus is not infrequently met with in persons who
never have given evidence at any period of having
suffered from catarrh, gouty or otherwise. That the
mucus of the urinary passages furnishes the colloid
medium by which the stone grows is undoubted ; but

I think in the case of the nucleus, the colloid medium
is furnished in some other way than the mucus of the
urinary passages. Some time since I suggested* that
instead of the calculous matter being originally
deposited in the pelvis of the kidney, the deposition
might primarily occur in the cells forming the wall of
the renal tubules as the result of some vital impairment, so that the products normally eliminated by
them were retained and deposited instead. In support
of this view I would urge the fact, that, in the urine
of those animals who secrete uric acid in a semi-solid
state, it is not difficult to find evidence of this substance being contained in the cells of the renal epithelium. Besides, the late Professor Quekett (*Medical
Times and Gazette*, 1851) gives the figure of a crystal
of calcium oxalate enclosed in a human renal cell. It is
more difficult, however, to prove that calculi are formed
in the tubules, since they must readily shell out and
pass into the pelvis of the kidney. But Prout, as the
result of his observations, came to the conclusion that
uric acid was secreted by the mammillary processes of
the kidney, in a semi-fluid state, which afterwards
becomes hard, and contracts as it hardens. Sir
Benjamin Brodie,† in confirmation of Prout, refers to
the fact that in Dr. William Hunter's Museum, which
was formerly in Windmill Street, but which is now in
Glasgow, there are several preparations illustrative of
this point in pathology ; to wit, the formation of
calculi at the mammillary processes. In some of these
preparations " the mammillary processes have been
longitudinally divided, and the tubuli uriniferi are
seen blocked up with calculous matter ; and in one of
them, the development of the calculus being further

* *Lancet*, vol. ii., 1883 ; and Quain's " Dictionary of Medicine,"
article, Calculus.
    † " Lectures on Diseases of the Urinary Organs," 4th edition,
p. 239. Longman & Co.

advanced, it is seen partly embedded in the apex of the mammillary process and partly projecting into the infundibulum."

Again, in the uric acid infarcts occurring in the tubules of young infants we find these to consist of small granules, spheroids, etc., an evidence that the deposition has taken place in the presence of a colloid ; and this colloid, I imagine, is more likely to be a renal cell, than furnished by a gouty, or specific catarrh. Those who admit the probable formation of the nuclei of calculi in the mammillary processes of the kidney attribute their formation to the more concentrated condition of the urine in the straight portion of the tubule than in the convoluted part. The degree of concentration is, however, very slight, whilst the increased diameter of the tubule in this part of its course, and there being less obstruction to the onward flow than in the convoluted part, would more than compensate for any increase of concentration that might occur. The true explanation of the occurrence of this deposit lies, I think, in the difference of the anatomical structure of this part of the tubule. If we examine a urinary tubule, we find it composed, from the neck of the capsule to the commencement of the ductus papillaries, of a wall of basement membrane (tunica propria) on which the epithelial cells lie; from the ductus papillaries to the apices of the pyramid the tubules lose their basement membrane, so that the wall is here formed of the epithelium alone, just as occurs in the ducts of the sweat glands where they perforate the epidermis. This portion of the tubule is also less freely supplied with blood than the other parts of the medulla. In the medulla the blood is conveyed by long vessels, the arteriolæ rectæ, which collectively enter the medulla from the side of the cortex. These arteriolæ rectæ proceed from branches of the renal artery, which also give off the arteria

inter-lobularis to the cortex.   The arteriolæ rectæ run
towards the fissure-like spaces in the marginal portion
of the medulla between the fasciculi of urinary tubules.
A brush of parallel vessels arises from the trunk of
each arteriolæ rectæ, and when the vessels of this
brush come into contact with the converging bundles
of the urinary tubules, they break up into capillaries
that form looped plexuses round the tubules ; and as,
on account of the progressive narrowing of the fissure,
one artery after another thus reaches the fasciculus of
the tubules, so do they successively break up into
capillaries.   The number of the arteriolæ consequently
diminish towards the papillæ, until in the latter
only one or two remain, which break up into capillaries and are distributed over the papillæ themselves.
It will thus be seen that less blood circulates through
this part of the kidney tubule, and we know by
analogy that textures possessed of fœble circulation
are particularly prone to undergo degenerative change
of some sort or other.   The fact also that the basement membrane disappears at this part of the tubule,
and that the wall consists alone of epithelium, may
also tend to produce degenerative changes in the cells
composing it.

A consideration of the clinical conditions under
which we find calculus adds some support to view I
have adduced.   Calculous disease is most frequent
during the period of childhood, early youth, and old
age,* periods when the tissues, either from rapidity of
growth or general impairment of vitality, are most
likely to undergo atrophic and degenerative changes.
After the age of puberty and during the period of
middle life calculi are comparatively rarely met
with, although attacks of gravel are common.   We
frequently find the history of a calculus associated
with some previous illness, in which the vital powers

* Statistics collected by Sir Henry Thompson.

have been much exhausted. Blows on the loins, violent strains of the back, etc., often lead to the formation of calculous matter in the kidney; undoubtedly in some of these cases extravasation of blood into the tubule may form the nucleus, but there are many cases where a most careful examination of the nucleus fails to prove a hæmic origin. The frequent association of stone with gouty tendencies may be explained by the existence of this state of impaired vitality and textural degeneration, and the calculous deposit that occurs in the renal tubules is caused in a similar manner as the deposits of sodium urate, in the parts poorly supplied with blood, as the cartilages of the joints and ear ; a deposit which is undoubtedly due to the inability of these tissues to eliminate the insoluble urate.

But of greater importance than the mode in which calculi are formed is the question of their disintegration, since it is upon a study of the conditions that tend to produce this result that we can hope to effect their removal from the body. Calculi in the renal or urinary organs break up, or disintegrate, in three ways : (1) *By fracture from direct violence.* Sir Benjamin Brodie (*op. cit.* 286, 287) has recorded instances of stones being fractured, apparently by concussion against each other. To allow of this, the calculous material must have been poor in organic matter and consequently very brittle ; or there may have been a friable layer interposed between two harder ones, as in Dr. Ord's case (*Path. Soc. Trans.*, vol. xxx., p. 319). (2) *Spontaneous fracture.* Dr. Ord has collected several examples of calculi undergoing fracture, apparently from forces contained within themselves. In two cases (*Path. Soc. Trans.*, vol. xxviii., p. 171 ; vol. xxix., p. 162) it seemed to be caused by expansion of the nuclei ; in a third (vol. xxxii., p. 304), the spores and mycelium of a fungus

Q

were found mixed with the débris, and Dr. Ord thinks this fungus growth was the cause of the breaking up of the calculus. Mr. Pearce Gould has drawn my attention to a calculus recently removed at the Middlesex Hospital, which contained, in its interior, purulent fluid ; one can readily see how changes in a fluid contained within any of the layers of a calculus might lead to its disintegration, either from disruption owing to decomposition of the fluid, causing the evolution of gas, and so, as it were, blowing up the calculus ; or else from the drying up of the fluid and the caving in of the walls surrounding the cavity. (3) *Disintegration by medical means.* In spite of Sir Henry Thompson's statement, that he "cannot find that any patient certified to have stone after sounding by a competent surgeon, after a course of any solvent being again sounded, or submitted to autopsy, was found free from stone," considerable success attended the administration of solvent remedies by the physicians of the last century for the relief of vesical calculi. Sir Henry Thompson seems to have overlooked the cases reported, by Stephen Hales, F.R.S., and David Hartley, F.R.S., of the four patients examined and treated in Guy's and St. ·George's Hospitals by surgeons of those institutions under the observation of the president and censors of the Royal College of Physicians with a view of determining the merit of the solvent treatment.* Certainly Cheselden, Nourse, and Sharp were competent surgeons, and not likely, in testing what was considered a quack medicine, to have omitted to state the fact if they had been able to detect a stone on the subsequent sitting. As these cases have been overlooked by modern writers on calculous diseases, I quote them in brief in order to show the success attained by this mode of treatment.

* Tracts B (No. 4) 250, in the Library Royal Medical and Chirurgical Society.

Case 1. Mr. Gardiner æt. 61. Sounded by Mr. Nourse, in presence of Mr. Wall, apothecary, Nov. 30, 1738. Stone detected. Took solvents for eight months, during which time he passed many fragments. Sounded again Sept. 14, 1739, by Mr. Sharp, in the presence of Mr. Cheselden, Mr. Sainthill, Mr. Belcher; no stone detected; relief from all symptoms.

Case 2. Peter Appleton, æt. 67. Sounded by Mr. Sharp, Guy's Hospital, July 6, 1739, in presence of Dr. Pellet, President; R. C. Phys. Dr. Whittaker, Censor; and Dr. Nesbit. A stone found which was considered a large one. Took solvents for five months, during which time he passed much grit. Sounded again Nov. 9, by Mr. Sharp, in the presence of thirteen surgeons and physicians; no stone could be detected; relief from all symptoms.

Case 3. Henry Norris, æt. 55. Sounded at St. George's Hospital, Aug. 17, 1739, by several surgeons. Stone detected. Took medicines four months, during which he voided a thick sediment. Sounded again at St. George's, Dec. 14; no stone could be detected; relief from all symptoms.

Case 4. William Brightly, æt. 79. Sounded at Guy's Hospital, Sept. 8, 1739, by Mr. Sharp and Mr. Gardiner. Stone detected. Took solvents for four months, voided grit freely. Sounded again at Guy's Hospital by Mr. Sharp; no stone could be found.

In addition to these well-authenticated cases, I have collected some 130 cases in which the use of solvent remedies was followed either by complete relief or diminution of the suffering. The introduction of lithotrity as an operation has rendered the use of solvents unnecessary as regards the treatment of vesical calculi; but the question may be asked why, if such successful results were obtained last century with solvents as regards vesical calculi, why do we not succeed with the same remedies in treating rena!

calculi now?   The answer is, that the conditions
under which renal concretions can be submitted
to solution are distinct from those which subsist
in the case of vesical concretions.   With the latter
we have the urinary bladder, capable of holding
four ounces of strongly alkaline fluid completely
surrounding the stone, and, moreover, easily kept
alkaline ; whilst, to increase the concentration of
fluid in the bladder, the patient was ordered to drink
sparingly, and to restrain himself from passing water as
long as he could.   With renal calculi, however, the
case is different.   The urine in contact with the
calculus at any given time is only a small quantity,
and as this passes away directly to the bladder, it is a
matter of extreme difficulty to keep the urine constantly
alkaline.   Alkaline remedies have, consequently, to be
given more frequently, and the aggregate amount of
alkali required to effect anything like a decided chemical action is much more than would be the case than
when the whole of the alkaline urine was collected in
the urinary bladder.   Moreover, there is the greatest
difficulty in maintaining hour by hour the urine in
an alkaline state.   A dose of alkali is speedily
eliminated, and though it renders the urine strongly
alkaline for a time, at the end of two hours the effect
has generally passed off, and the urine becomes acid.
So while the urine in the bladder may remain alkaline from the first portion of the secretion, the latter
portions as they come from the kidney may be acid.
Moreover, the experiments of Bence Jones, Parkes,
Beneke, and myself,* have shown that the alkaline
bicarbonates, whilst they render the urine alkaline for a time, subsequently tend to increase the
acidity.   And it is probable the vegetable salines,
the citrates and tartrates, act similarly, since they are
reduced to the condition of bicarbonates in the blood.

* *Lancet,* Nov. 9th, 1878.

At all events, Dr. Bence Jones found that although five drachms of tartrate of ammonia were taken in one day the urine was not made alkaline, whilst with a fixed alkali like potash this large dose was required to keep the urine alkaline for a portion of the day only. Not only also is there a difficulty in getting patients to take these large doses with sufficient frequency and regularity, but I doubt the propriety of giving such large doses of the alkalies for long periods of time. I have found quite as much good result from the continued use of distilled or soft water. Dr. Murray, a few years back, called attention to this mode of treatment, and I have since employed it in all cases coming under my notice, and I can fully confirm all Dr. Murray said in its favour. In one case (*Path. Soc. Trans.*, vol. xxxiii., p. 206) the calculus was passed as a mere shell, much eroded, after two years' persistence in the use of soft water, just at a time when the question of operative procedure was being entertained. In cases of smaller calculi they came away more freely, their surface being eroded, showing that some diminution in bulk had been occasioned. The action of distilled water in promoting the disintegration of calculi may be explained by its taking up inorganic matter and withdrawing it from the body ; whilst by taking four pints daily of soft water instead of four pints of ordinary drinking water, about 80 grains of saline constituents (chiefly lime salts) are cut off. It is evident, therefore, if we diminish the supply of inorganic matter, less will be withdrawn from the body, so that after a time the urine will contain less and less ; and it has been shown that the poorer calculi are in inorganic constituents, the greater their tendency is to disintegrate. Again, soft water has a powerful diuretic action, increasing the urinary flow by 12 to 20 per cent. and diminishing the specific gravity. Now, Mr. Rainey in his work on molecular coalescence has shown that bodies placed in solutions of different density, to that in which they were formed, have a tendency to undergo molecular disintegration. Lastly, soft water seems to diminish the catarrh of the urinary passages, and thus prevents the excessive formation of mucus, which aids in the growth of the calculus. This is a point that ought to be more attended to than it is, for not only does the reduction of the catarrh prevent any further increase in growth, but it gets the urinary passages into a better state to allow the passage of the stone.

Chemical examination of urinary calculi. — Notice the size and general appearance, whether in section they are made up of concentric layers, or present a uniform surface. A portion of the stone is then broken down and reduced to powder. If the stone is made up of different layers, a portion of the powder of each layer must be submitted to analysis.

CLASS I. Little or no residue is left when a small portion of the powder is burnt on platinum foil under the blow-pipe. The calculus may consist of Uric acid, Xauthin, Cystin, or Hæmic elements, or is Fibrinous, or composed of Urostealith.

(1) *Uric acid calculi* are the most common of all calculi, constituting about 80 per cent. of all varieties, and often attain a considerable size; they are usually smooth, and of a light yellow or reddish-brown colour. The burnt powder evolves an odour of prussic acid and of burnt animal matter; the ash is extremely small, and contains traces of sodium phosphate and carbonate. Dissolved in liquor potassæ, and reprecipitated by the addition of hydrochloric acid, the characteristic crystals will be observed under the microscope. Heated with nitric acid and touched with ammonia the purple colour (murexide) will be produced.

(2) *Xanthin calculi* are extremely rare; they are usually smooth, and of a cinnamon colour, taking a polish when rubbed. They dissolve in nitric acid without effervescence; and do not yield the murexide reaction with nitric acid and ammonia. To obtain crystals of xanthin dissolve a little of the powder in hydrochloric acid on a glass slide, and allow the solution to evaporate slowly.

(3) *Cystin calculi* are also very rare, they have a smooth surface, a greenish-yellow colour, and break with a crystalline fracture; they are the softest of all calculi, and for some days after their removal are compressible. Cystin dissolves in the caustic alkalis and in strong mineral acids; on evaporating the ammonia solution cystin is deposited in regular hexagonal tables; on evaporating the hydrochloric solution cystin crystallises in radiating needles. Boiled in a solution of caustic potash with a little lead acetate, a black precipitate of lead sulphide is thrown down, which is due to the presence of sulphur contained in the cystin.

(4) *Hæmic, fibrinous, urostealith*, and *indigo concretions* clear considerably, and leave little or no ash. The hæmic concretions both microscopically and chemically show that they are derivations from blood. The fibrinous concretions give the reaction of fibrin. The urostealith concretions, which are a mixture of fatty and soapy matters, mucus and withered cell forms, dissolve in ether. Concretions of indigo are sometimes met with. (*See* Dr. Ord's case, *Path. Soc. Trans.*, vol. xxix., p. 155.) The best test for the presence of indigo is the reducing action of glucose in an alkaline solution (§ 119, page 150). The probable explanation of the deposit of indigo in the urine will be found § 77, page 53.

CLASS II. A considerable residue is left when a small portion of the powder is burnt on platinum foil. The calculus may consist of Calcium phosphate, Ammonio-magnesium phosphate, Calcium oxalate, Calcium carbonate, or Urates.

(1) *Calcium phosphate.*—Calculi composed solely of this substance are extremely rare ; when met with they are often of considerable size. They have a white chalky appearance, and their surface is very friable. Most commonly calcium phosphate is mixed with ammonio-magnesium phosphate, forming the mixed or *fusible* calculus (*q.v.*). The ash does not fuse (*infusible*) under the blow-pipe, is soluble without effervescence in hydrochloric acid. The acid solution gives a precipitate with ammonium oxalate, indicating the presence of lime ; and a yellow precipitate with uranium nitrate, shows the presence of phosphoric acid.

(2) *Ammonio - magnesium phosphate*, rare as an entire concretion, but frequent as a layer or crust of other calculi. Since it is formed whenever the urine becomes alkaline from the presence of volatile alkali (ammonia), the result of ureal decomposition set up by morbid state of the urinary passages, and thus it becomes a frequent component of urinary calculi. Under the blow-pipe the ash fuses with difficulty, and dissolves in hydrochloric acid without effervescence. The acid solution, when ammonia is added in excess, throws down the characteristic crystals of triple phosphate.

(3) *Fusible calculus*, a mixture of calcium phosphate and triple phosphate, often with some addition of calcium oxalate ; frequently attain a considerable size in a short time. Under the blow-pipe *fuses* into a glistening enamel, which adheres firmly to the platinum foil on which it is fused. The powder of the calculus dissolves readily in hydrochloric acid ; the acid solution, on the addition of ammonium oxalate, throws down a precipitate of calcium oxalate, indicating the presence of lime. If this is filtered off and ammonia added, crystals of triple phosphate will crystallise out.

If the calculus contains traces of calcium oxalate, the ash will dissolve with effervescence, and a portion of the calculus itself will not dissolve in acetic acid.

(4) *Calcium oxalate calculi* are either small and pale-coloured, or large, dark-coloured, with a rough, irregular surface, and on section present an angular structure, with irregular, dark-coloured laminæ. The smaller calculi, from their size and appearance, are termed "hemp-seed calculi;" the larger "mulberry calculi." Very rarely it is deposited on other calculi in a crystalline form. (*See* case reported by Mr. Morrant Baker, *Path. Soc. Trans.*, vol. xxv., p. 261.) Heated on platinum foil, it first of all chars from the combustion of the organic matter, and gives off an odour of burnt animal matter; finally the residue becomes white, and consists of calcium carbonate, which dissolves with effervescence on the addition of hydrochloric acid, owing to the reduction of the oxalate to a carbonate by combustion. This solution, when neutralised with ammonia, throws down a precipitate of calcium oxalate on the addition of oxalic acid. A portion of the calculus itself is insoluble in acetic acid.

(5) *Calcium carbonate calculi* are rare, but when met with are generally multiple, occurring in large numbers, often derived from the prostate gland. They are spherical in shape, sometimes cubical or pyramidal. On section they appear to be composed of concentric rings, and by polarised light they sometimes display a dark-looking cross. Powdered, they dissolve in hydrochloric acid with effervescence. Under the blow-pipe the powder slowly fuses.

(6) *Urates.*—Calculi composed entirely of urates are very rare; they are generally found mixed with uric acid. They are distinguished from this body by the powder being soluble in hot water, which dissolves the urates, leaving the uric acid. The filtrate evaporated gives with nitric acid and ammonia the murexide reaction.

153. **Biliary calculi** vary in size from mere grains to masses as large as pigeon's eggs, or even larger, since stones measuring from two to two and a half inches long, and one inch thick, are to be found in most hospital museums. Their number is in inverse proportion to their size; the smaller they are the more numerous. Some thousands of small calculi, varying in size from a grain of sand to that of a small pea, are often recorded as having been removed from the gall bladder. The medium-sized stones are generally multiple, but they are not present in such excessive numbers, whilst the large stones are usually solitary. It is rare for one stone to differ from its fellows, taken from the same gall bladder. However numerous the calculi may be, or variable in size, in external appearance and chemical composition they will be found to correspond. Gall stones, though more or less rounded form, vary considerably in shape; thus they may be cubical, pyramidal, polyhedral, etc., with the edges rounded off and facetted, or their plane surfaces rendered concave. These changes from the rounded or ovoid form are brought about by mutual pressure, when there are more than one calculi present in the gall bladder. With solitary calculi the rounded or ovoid form is generally preserved. Foliated and branched concretions are rarities. In external appearance they are generally smooth and slightly greasy to the touch. Others have roughened, nodulated surfaces, resembling in appearance lychee nuts. The colour varies from a dirty white to a yellow-brown, or even deep black colour. The prevailing shades, however, are brownish black (sepia), deep olive-green, and russet brown. In consistence they are soft, except in some forms, when the crust consists largely of lime salts. This portion of the calculus is often unusually hard. On section biliary calculi present a variety of appearances. They may have a simple uniform structure, apparently homogeneous

throughout, breaking with either a crystalline or
an earthy fracture ; these calculi are, however, comparatively rare. The most common forms are those
that present a distinct nucleus, a body, and sometimes
a crust or thin rind covering the body. The nucleus, in
recent calculi, is generally homogeneous in appearance,
but in old and dry gall-stones it is often shrivelled and
fissured. It consists for the most part of inspissated
mucus, with bile pigment and cholesterin ; sometimes
it consists almost entirely of cholesterin. Foreign
bodies often form the nuclei. Thus, round worms,
distoma, fragments of needles, plum stones, aggre-
gations of quicksilver, have all been met with in the
interior of biliary calculi. The body of the calculus
on section presents (*a*) an amorphous appearance of
brownish-yellow colour, the material being arranged in
concentric layers round the central nucleus, the external
surface having sometimes a crust or rind, but oftener
no distinct crust or rind can be made out. (*b*) From
the nucleus long stratified crystals of cholesterin
radiate towards the circumference of the stone, which
is covered with a dense and often nodulated crust.
Various explanations have been offered to explain why
some calculi are amorphous and others crystalline.
Schueppel thought that it was the mingling of
the bile colouring matter with the cholesterin that
prevented it separating in the crystalline form.
Wherever, then, according to this view, there is
much pigment, the calculi are amorphous. Dr. Ord
(*Path. Soc. Trans.*, vol. xxxi., p. 141) explains it
thus : When first precipitated the cholesterin is dis-
seminated through a bed of biliary pigment, a colloid
of high molecule. This not only prevents the formation of separate crystals, but tends to group the
crystalline matter into spherules. If the colloid always
remained a colloid, no further change would occur; but
the colloid pigment tends in process of time to become

crystalline, and as step by step it assumes that condition, the two substances are segregated into two
zones, a central one of the more adhesive pigment,
an outer one of the polar crystal. An experiment of
Dr. Ord's, showing the separation of cholesterin from
bile pigment, after deposition in combination there-
with, may be cited in illustration of the proposition
advanced. A gall stone containing a great deal of
pigment is reduced to fine powder. Some of this is
mixed on a glass slip with glycerin and glacial acetic
acid, and the mixture is covered with thin glass. The
slip is then slowly heated, over a spirit flame, to
ebullition. When the powder is in great part dissolved, the slip is transferred to the microscope. The
fluid is found in a state of great agitation, and filled
with very small yellowish spherules, which run
together to form large spherules of all sizes. These
move about the field under the influence of the
currents, with all sorts of amœba-like changes of outline ; but they are perfectly homogeneous, and do not
affect polarised light. Presently they become station-
ary, lose their transparency, and begin to crystallise,
the process beginning at one point and extending
thence quickly over the whole mass. The crystallisation converts them into lozenge-like bodies, covered
with somewhat round or angular projections, finely
marked with bent parallel lines indicating imprisoned
acicular crystals. Very often the interior remains
uncrystallised, but in a state of evident tension for
some time. The crystals are perfectly colourless, the
pigment being separated from them in subcrystalline
nodules of a brilliant yellowish-red, like that of hæmatoidin. After a varying time, sometimes many hours,
the imprisoned raphides burst from their envelope,
and the whole mass bristles with them, star-fashion.
At the same time the tense interior arranges itself in
concentric laminæ of parallel radiating crystalline

fibres, still completely separate from the pigment, which forms alternate layers ; so that we have before our eyes, within a few hours, the spectacle of the formation of a tiny biliary calculus.

In this experiment, however, the solution of cholesterin and bile pigment was first heated, and it is natural to expect that, in cooling, the least soluble of the two would separate out first, and that would be the cholesterin ; but it is doubtful whether the conditions for such separation exist in the body, unless the cholesterin is in considerable excess.   I believe the whole question depends simply on the amount of mucus present in solution.   If this is in excess, then the cholesterin separates out slowly in an amorphous form, carrying with it the pigment.   If the mucus is not abundant, and the cholesterin is in excess, it is deposited rapidly in a semi-crystalline state, leaving the pigment behind. A chemical examination of these calculi shows that after the cholesterin has been removed by ether there is very little organic residue, whilst the crust is poor in pigment but rich in lime salts, which sometimes assume a rod-like spiculated form.   All gall stones contain cholesterin.   Some, indeed, consist almost entirely of this substance, these concretions, which are not frequent, being generally met with in young children.   The mixed calculi are by far the most common, containing a variable proportion of cholesterin, from 20 to 90 per cent., mixed with bile pigment, traces of a fatty acid in combination with lime (margarate of lime; Frerichs), and sometimes a small quantity of bile acids, and an organic matrix derived from the mucus of the gall bladder.   The inorganic residue consists chiefly of lime salts, carbonates and phosphates, traces of iron, and sometimes copper.

An analysis of gall stones is conducted as follows : Separate portions of the crust, the body, and the nucleus, are to be submitted for analysis.   Weigh a fragment, incinerate under the blow-pipe, weigh the ash ; the difference represents the amount of organic matter present ; examine for lime salts (§ 84). Thoroughly exhaust another portion with ether, decant off the etherial solution and evaporate; weigh; this gives the amount of cholesterin present. Test for cholesterin: by adding a drop of nitric acid, heating, and touching residue with ammonia, when a reddish-brown coloration will be given ; or by evaporating with ferric chloride, and touching with hydrochloric acid, when a violet blue is yielded. The residue left after exhaustion with ether is then treated with chloroform, and the chloroformic solution evaporated, when a brownish powder of bile pigment will be obtained (§ 140). Some calculi consist almost entirely of pigment. They are rare, however, are small like gravel, and have a tarry blackish lustre. When broken across they appear homogeneous.

154. **Pancreatic calculi** are of rare occurrence, usually associated with an atrophied condition of the gland. They are generally multiple, and are found in the main and accessory duct of the organ. They are oval in shape, and their surface often presents a worm-eaten appearance of whitish colour, which when rubbed acquires an enamel-like lustre. When broken across, the fracture presents a glistening white porcelain appearance. In some calculi I reported on for the Chemical Committee of the *Pathological Society* (Trans., vol. xxiv., p. 137) I found they consisted of 24 per cent. organic matter, and 76 per cent. inorganic ash. The latter chiefly consisted of calcium carbonate, calcium phosphate being in smaller proportions ; the soluble salts, chlorides and carbonates of potash and soda, being extremely small. Process of analysis: Weigh a portion of the calculus, reduce it to fine powder, incinerate, weigh ; the difference in weight gives the amount of organic matter and water. Divide

the ash into two portions. With one, estimate amount of lime (§ 84), potash, and soda (§ 88). With the other, dissolve a part in acetic acid, and estimate the phosphates (§ 113); and the other part in hot water to determine the chlorides (§ 114). The carbonates are determined as follows : A weighed portion of the ash is dissolved in water and introduced into a small flask ; this flask is fitted with a bulb tube filled with dilute nitric acid, which is prevented from flowing into the mixture by means of a pinchcock. The apparatus is now weighed and attached to another flask containing concentrated sulphuric acid, and the bulb tube pushed down to nearly the bottom of the flask, the pinchcock pressed and the dilute nitric acid allowed to mix with the contents of the flask. When the carbonate is completely decomposed, the flask is placed in warm water and gentle suction applied to a tube passing through a perforated cork of the second flask till all the carbonic acid is removed ; the apparatus is then allowed to cool and is again weighed; the loss of weight represents the amount of carbonic acid present in the salt examined. The amount of carbonic acid thus obtained is, however, somewhat in excess of that existing in the form of carbonate in the concretion, since some of it is derived from the combustion of the organic matter.

155. **Intestinal concretions** are found chiefly in the cæcum and large intestines. They vary considerably in size and composition. In colour they are yellowish, inclining to grey or brown tints. Their nucleus is usually a foreign body, a gall-stone, woody fibre, fruit stone, etc. Intestinal concretions having a definite composition consist chiefly of ammonio-magnesium phosphate, calcium phosphate, carbonate and sulphate, organic matter and fat. The analysis of these calculi is to be conducted as for urinary calculi (Class II.). They should, however, always be extracted with ether, to see if they contain cholesterin. When carbonate of

magnesia has been taken habitually it may accumulate
and concrete in the intestines. Concretions of yellow
waxy appearance are sometimes passed, varying in
size from a pea to a filbert; they chiefly consist of
fatty matter from 60 to 70 per cent., mixed with
earthy phosphates (lime and magnesia) and an
animal substance of fibrinous nature. The fatty
matters can be removed by ether, and the etherial
solution examined to ascertain the nature of the fats
as directed (§ 99, page 85). The organic residue or
fibrinous mass will give the reaction with hydrogen
peroxide (§ 23); the earthy phosphates estimated as
directed (§ 113, page 136); the lime and magnesia as
directed (§§ 84, 85, pages 59, 60). Other concretions
may consist of masses of hair, woody fibres, and husks
of seeds. In Lancashire and Scotland concretions
composed of the caryopsis and fragments of the envelopes of the oat, studded or encrusted with crystals
of triple phosphate, are not uncommon. Concretions of this character are only to be detected by a
microscopical examination. Animals, especially the
herbivora, suffer much from intestinal concretions.
In Persia and Thibet, peculiar stones termed " bezoar"
are met with in a species of antelope. They contain
no cholesterin, and yield only little ash when burnt,
and seem chiefly to consist of ellagic acid, an acid
related to gallic acid. It is, therefore, probable that
these concretions are derived from a vegetable source
in the articles used for diet.

156. **Salivary calculi** are of rare occurrence in
man. They generally occur in Wharton's duct near
its outlet, where they may be easily recognised. More
rarely they are situated deep in the main duct near
the gland (*Path. Soc. Trans.*, vol. xxiv., p. 88). The
nucleus often consists of a splinter of wood or a fragment of bone. They are generally of rounded oval
form, greyish-yellow in colour, and tuberculated on

the surface. Their average composition is: Organic matter and Water, 13; calcium carbonate, 80; calcium phosphate, 4; and magnesium phosphate and carbonate, 3 per cent. For analysis proceed as directed for pancreatic calculi.

157. **Prostatic concretions** occasionally form in the body of the prostate gland. They are of two kinds, (*a*) very small rough concretions varying in size from a poppy seed to a mustard seed, extremely numerous, of yellowish colour, occurring in the gland before any extensive disorganisation takes place ; (*b*) larger concretions, of irregular and porcelainous appearance, generally met with when the gland is greatly disintegrated. The former variety consists of calcium carbonate, mixed with a little calcium phosphate ; on section they exhibit a series of concentric lines. Their powder effervesces strongly on the addition of hydrochloric acid, and yields an abundance of carbonic acid gas. The acid solution, with ammonium oxalate added in excess, throws down a precipitate of calcium oxalate, indicating the presence of lime ; whilst a few drops of uranium nitrate solution when added to the acid solution throws down a precipitate showing the presence of phosphoric acid. The larger concretions contain less carbonate but more phosphate of lime than the smaller ones. It must not be forgotten that there is always a danger of some of the smaller calculi finding their way into the bladder, and thus becoming the nuclei of vesical calculi.

158. **Gouty concretions. Tophi.**—Gouty persons are subject to deposits of sodium urate encrusting the surfaces of the bones, infiltrating within tendons and their sheaths, and depositing in the cartilages of the ear. In the majority of cases these deposits lead to marked deformities, but in others, though not a trace of external deposit may be perceptible, it can be demonstrated that the cartilages and other

R

structures of the joints may be infiltrated with sodium urate. Indeed, Dr. Garrod insists that gouty inflammation is invariably attended with the deposition of sodium urate. Certainly the deposit is never met with in acute or chronic forms of rheumatism, nor in rheumatoid arthritis; but whether an attack of gout invariably leads to the deposition of sodium urate is a point that requires further evidence before it is finally settled, though the balance of evidence is certainly in favour that it is. The cases brought forward hitherto to disprove it were not examined so completely as they ought to have been, and minute traces are easily overlooked. Before deciding that no sodium urate is deposited, a most searching investigation is required of the cartilage and synovial membranes, since, when present in extremely small quantities, it is likely to escape observation. When first deposited the sodium urate is in a semifluid state, but it soon becomes hardened and chalk-like. On making a vertical section of cartilage affected with this deposit, it will be found that it does not extend very deeply, rarely exceeding two-thirds of its depth. The deposit appears to the eye amorphous, but on microscopic examination it is found to consist of small granules mixed with crystalline needles. If the cartilage be cut into thin slices and washed with cold water and alcohol, and then digested in hot water, the deposit is removed, and the cartilage becomes transparent. The aqueous solution being evaporated, crystals of sodium urate will be deposited in small tufts. Evaporated with nitric acid and the residue touched with ammonia, they give the purple (murexide) reaction of uric acid, whilst the ash left after incineration with the blow-pipe gives abundant evidence of the presence of soda. These deposits are said to have been found in other situations besides those above enumerated, as in the lungs, the bronchial tubes, the

meninges of the brain, and in the concretions on the
valves of the heart and in the atheromatous deposits
in the aorta.   Dr. Garrod, however, has failed to find
the least trace of uric acid in such situations, and
suggests that tabular crystals of cholesterin may have
been mistaken for that body.   I would further suggest
that the chemical test, if applied, might mislead, since
cholesterin evaporated with nitric acid and touched
with ammonia gives a brownish-red coloration that
might be mistaken for the purplish red given by
uric acid with this test.   In the gouty kidney
white points and streaks are to be seen upon the
pyramidal portion of the kidney.   In these cases the
sodium urate is generally believed to be deposited in
the tubules, but Dr. Garrod thinks that it is embedded
in the fibrous structure itself.   The microscopical
appearances of this condition of kidney are admirably
described by Dr. George Johnson, in his work on
diseases of the kidney.

159. **Miscellaneous concretions.**—These consist chiefly of varying quantities of calcium carbonate
and phosphate, with much fatty matter, chiefly cholesterin, fibrin, casein, gelatin, etc., and proteid bodies.
Such are the pulmonary concretions, deposits in muscular tissues, nasal concretions, etc.   The following are
a few of the recorded analyses giving the percentage
amount of the organic matter and calcium phosphate
and carbonate :

	Solid tumour from uterus.	Ossified muscle.	Concretion from brain.	Osteoid tumour from lungs.	Nasal concretion.
Organic matter and water	36	58	22·46	38·89	5
Calcium phosphate	56	32·09	60·32	53·33	90
Calcium carbonate	5	8·66	17·21	7·04	5

In a concretion taken from the brain of a horse,

Lassaigne found 58 parts of cholesterin, 39·5 proteid matter, and 2·5 of calcium phosphate (Simon).

160. **Products of degeneration.**—These are generally divided into two classes. Those in which there is a direct *metamorphosis* of the proteid elements of a tissue, and which is generally followed by the destruction of the histological elements, and the softening of the intercellular substance, so that the composition and structure of the tissue may be completely altered. And those in which there is no conversion of the proteid elements themselves, into a material of another kind, but the new substance is introduced from without by the blood by a process of *infiltration*. Here the cell elements and intercellular substance are not softened or destroyed to any great extent, and even then it is due to secondary metamorphosis produced by the infiltration, as the fatty changes that are associated with lardaceous infiltration. 1. *Fatty degeneration* may occur either as an infiltration or a metamorphosis, but no decided line of demarcation can be often made between them, since the same cause may be in action at the same time, both within and without the tissues ; and also, the deposit of fat, having accumulated, may, in consequence of the pressure it occasions, lead to fatty metamorphosis as well. In fatty infiltration, when the process is distinct, the fat is deposited in oily drops within the cells. These at first are small and distinct, but as the process proceeds they accumulate and fill the cell, obscuring the muscles and protoplasm. But these remain unaltered, so that if the fat is removed they may be restored to their original condition. Fatty infiltration occurs chiefly in the voluntary muscles, which have been disused, as in paralysis, joint diseases, lead palsy, etc. ; in the muscular tissue of the heart, and especially in the liver. In this latter instance it is important to distinguish it from the acute fatty changes

that are brought about by the action of certain poisons, etc. In fatty liver, due to infiltration, the organ is increased in size, the surface is smooth, and the edges rounded. The section presents an opaque yellowish-white colour, in early stages only affecting the portal area of the lobules, but gradually extending to the whole lobule, leaving only a reddish point in the centre, at the origin of the hepatic vein. Fatty infiltration is a chronic process. It may be induced by excessive use of fatty food and alcohol, and is the frequent result of chronic wasting diseases, in which the oxygenating power of the blood is diminished. In acute fatty degeneration of the liver the organ is diminished in bulk, the surface is wrinkled, and the edge sharp. On section it presents a yellowish-red colour, which may be uniform throughout, or else mixed with patches of brick-red (Zenker's patches). The lobules have entirely disappeared, and the cells consist only of fat granules and débris. The process is acute. It is met with in the disease known as acute yellow atrophy, in phosphorus poisoning, after the injection of the bile acids, and in some few instances it has been found in patients dying of acute diabetic coma. Fatty infiltration is brought about by excess of fat in the blood; this may be simply as a consequence of general obesity, or from defective oxydation, which accounts for the fatty changes which are associated with chronic pulmonary disease, especially phthisis. Fatty degeneration differs from the preceding, in that the fat is not derived directly from the blood, but by retrograde changes taking place in the proteid constituents of the tissues themselves. Extremely minute granules of dark colour of strong refractive power, and very soluble in ether, first diffuse through the protoplasm. As the process continues, the nucleus is invaded and the cell wall disappears. The granules at first are more or less coherent, forming rounded masses (corpuscles of

Gluge); but after a time these break down and the
granules are distributed through the tissue. As a
result of fatty degeneration the tissue may be entirely
destroyed, or the fatty product may be absorbed, or
it may undergo further changes and become caseous.
Fatty degeneration may attack any tissue or organ.
In the arteries and capillaries, as a primary condition, fatty degeneration is a senile change; but
it is also secondary to inflammatory conditions of
the vessels, in which the deposit of fat is preceded
by a cellular infiltration of the subendothelial connective tissue. Fatty degeneration of the heart
may be diffused or localised. The former may occur
in the course of those diseases in which oxydation is
reduced to a minimum, as in anæmia. Dr. Green
(*Trans. Clin. Society*, vol. viii., 1875) instances a case
of acute fatty degeneration of the heart, induced
by profuse loss of blood during the menstrual period
and inability to take food. In a minor degree fatty
degeneration of heart occurs in most pyrexial conditions. Localised fatty degeneration of the heart
is generally the result of disease or obstruction of
the coronary arteries, or the effect of pericardial inflammation. Fatty degeneration of the brain may
be either acute or chronic. In the former it is generally due to sudden cutting off of the blood supply,
as for instance by embolism of the middle cerebral
artery. In chronic softening, the supply of blood is
gradually diminished or slowly cut off, as occurs
when the vessels are diseased. In fatty degeneration
of the brain a considerable amount of glycerin-phosphoric acid is formed among the products of disintegration, occasioned by the breaking up of the
phosphorised fats, lecithin, etc. Acute fatty degeneration of the liver has been already alluded to. The epithelium of the kidney undergoes fatty changes after attacks of inflammation, as in acute and chronic nephritis,

and also in acute yellow atrophy, phosphorus and sulphuric acid poisoning, and in some cases of diabetes. More fat than ordinary is found in the epithelium of persons dying from chronic wasting diseases. In order to investigate the nature and degree of the fatty changes in any given tissue we first determine the amount of fatty matter present, the neutral fats, the cholesterin and lecithin, and then make a separate estimation of the neutral fats to ascertain the proportionate amounts of solid and oily fat. For the first determination, a weighed portion of the tissue, finely divided, is to be dried till it ceases to lose weight. It is then to be reduced to powder, and boiled with ether for some hours. For this purpose Drechsel's fat exhaustor is undoubtedly the best form of apparatus, but when not obtainable the following method can be satisfactorily employed. The finely-powdered tissue is introduced into a glass flask and half filled with absolute ether. The mouth of the flask is then fitted with a perforated cork, through which a fine glass tube, about two feet long, open at both ends, is introduced, the lower end of which should only be allowed to project just below the cork into the flask. The apparatus is suspended by means of a support-stand over a porcelain basin, so that about one-third of the flask is submerged, and filled with hot water about twenty degrees below boiling point. Over the upper third of the flask, and about half way up the tube, wrap a cloth dipped in cold water, as cold as possible. This prevents the too rapid evaporation of the ether, a considerable portion of which condenses and falls back into the flask. The temperature of the water in the basin is to be maintained by the constant addition of fresh supplies, as it is advisable not to bring a flame near the flask containing the ether. After the ether has boiled some considerable time, the cork with the tube is withdrawn, and the etherial solution, whilst still warm, is poured into a weighed platinum dish and evaporated. In order to remove all traces of fatty matter from the flask, it should be rinsed with a little absolute ether, and the rinsings added to the solution in the platinum dish. This is then to be weighed, and the increase of the weight gives the total amount of fat. In order to estimate the neutral fats, cholesterin and lecithin, separately, proceed as directed § 99, page 85. If it is desired to ascertain the proportion of solid fats to the oily, it is necessary to dissolve the fatty residue thoroughly with boiling alcohol, and filter. Boil the filtrate with a solution of potassium hydrate, which saponifies the fatty matter. The mixture is then evaporated, and the dry residue dissolved in water, which is acidulated with a little hydrochloric acid. On cooling, the stearic, palmitic, and oleic acids will be deposited. These are to be dissolved in boiling alcohol, and some alcoholic solution of lead acetate added. The precipitate, which consists of stearate, palmitate, and oleate of lead, is removed by filtration. The precipitate is treated with hot ether (*a*), which removes the oleate of lead, whilst the filtrate (*b*) contains the stearate and palmitate. Both the filtrate *b* and the etherial solution *a* are to be respectively treated with hydrochloric acid, and the mixture warmed ; the lead is then to be removed by sulphydric acid, and the precipitate removed by filtration. The clear filtrates are to be concentrated, and then shaken with ether, and the etherial solution allowed to evaporate, when from *a* oleic acid will be deposited, and from *b* a mixture of stearic and palmitic acids (the solid fatty acids) will crystallise out.

(2) *Lardaceous degeneration*, or, as it was formerly termed, amyloid degeneration, from its supposed relation to starch and cellulose. The morbid material is deposited in the first instance in the small arteries

and capillaries, chiefly of the liver, kidneys, spleen, and intestines, though all organs and tissues may be invaded. The cells of the intima are the first to be infiltrated ; thence it spreads to the muscular wall of the vessel, and from thence it passes to the cells and intercellular substance, till the whole organ or tissue is involved. It is still a question among pathologists whether lardaceous disease is a true degeneration (that is, a chemical and structural change in living tissues), or whether it is an infiltration among the tissues of a morbid material derived from the blood. On this question, the histologists generally support the view that it is an infiltration, the pathological chemist that it is a true degeneration of tissue. When first discovered, it was thought to be analogous to starchy matter, and was named amyloid ; but it was soon recognised that it was a definite nitrogenous body, and related to the proteid group, and the term lardacein was applied to it, to express the waxy or bacony appearance given to tissues affected by it. With regard to the nature of this substance various views have been expressed. Dr. Dickinson regards it as de-alkalised fibrin ; since he has found the tissues involved by it remarkably deficient in potash, and has prepared it artificially, by digesting fibrin in dilute hydrochloric acid, a substance which gives one of the reactions supposed to be characteristic of lardacein. In objection to Dr. Dickinson's views, it may be urged that, by digesting fibrin in dilute hydrochloric acid, syntonin, and not lardacein, is formed ; whilst the reaction with iodine, as I showed at the Pathological Society (*Trans.*, vol. xxx., p. 536), is not distinctive of lardacein, but may be given equally with dry fibrin, with syntonin, and with casein. This is important, since the red colour, being developed equally with alkali albumin (casein) as with acid albumin (syntonin), shows that the reaction is not caused by the

removal of an alkali. With regard to the deficiency
of potash existing in the affected tissues, that may be
accounted for by the great increase of fat found in
them ; and we know that tissues that have undergone
fatty degeneration become poorer in saline constituents.
The view I am inclined to adopt is, that lardacein is
a mixture of a proteid with a fatty body, of which we
have a physiological example in vitellin, and which
Hoppe Seyler considers to be a mixture of globulin
and lecithin. Lardacein may be separated from the
tissues by Kühne's process. The organ is finely
minced, and extracted repeatedly with cold water, and
subsequently with dilute alcohol, till the fragments become colourless. They are then digested with artificial
gastric juice. This has no action on lardacein, but
dissolves all the other proteids, leaving the lardacein
almost pure, with the exception of small quantities of
elastic tissue and mucin. Lardacein thus obtained
is soluble in dilute ammonia, from which it can be
precipitated by dilute acids. It is insoluble in water,
and does not perceptibly swell in solutions of sodium
chloride. Strong hydrochloric acid converts it into
acid albumin, and caustic alkalies into alkali albumin.
Touched with an aqueous solution of iodine, it
acquires a brown-mahogany colour, whilst methyl
aniline gives a rosy-red. Formerly, the blue colour
produced by the joint action of iodine and sulphuric
acid when poured on it was relied on as a test ; but
this blue coloration was shown by Dr. John Williams
(*Path. Soc. Trans.*, vol. xxvii., p. 322) could be
caused by the admixture of iodine and sulphuric acid
alone, without the intervention of any other substance.

Hyaline degeneration.—At the debate on lardaceous disease at the Pathological Society, 1879, Dr.
Stephen Mackenzie, with reference to the influence of
fever in producing the disease, called attention to a

change in the blood-vessels that occurred in certain
diseases with blood alterations. He had found it in
the arteries of the spleen in nearly all cases of pyæmia,
in some cases of diabetes, in a case of angina Ludo-
vici. Dr. Klein had first described the change in
typhoid and scarlet fevers. In most cases the change
was confined to the intima, but in some it involved
the muscularis. It consisted in a hyaline transformation, which did not give a characteristic reaction with
iodine, but stained slightly with methyl aniline. Dr.
Mackenzie did not regard the change as lardaceous,
but thought an examination into the nature of the
change might throw some light upon that disease. In
commenting on Dr. Mackenzie's remarks, I suggested
that the difference between hyaline and waxy degeneration might be only one of degree, the hyaline
being the first step in the degenerative process. A
microscopic examination of the hyaline substance
shows that, at first, it consists of granules of an albuminous nature, which are insoluble in ether; but, in a
more advanced stage, the granules are distinctly fatty
and soluble in ether. From this, it would appear
that the change is connected with fatty degeneration.
It may be, therefore, that hyaline changes are the
first step in a process which, if acute, leads to fatty
degeneration, if chronic, to waxy or lardaceous
disease.

The amyloid bodies, *corpora amylacea* of Kolliker
and Purkinje, are rounded oval bodies, varying in
size from extremely minute granules, which refract
light strongly, to bodies about one or two lines in
diameter, formed by the conglomeration of smaller
granules, apparently in a succession of concentric
layers. They are most frequently found in the bodies
of aged persons, chiefly affecting the prostate, the
ependyma of the ventricles, the fornix, the choroid
plexus, the retina, and spinal cord. They are apt to

become calcified, and in this form, when met with in
the nerve-centres, constitute the so-called "brain
sand." The corpora amylacea usually stain a deep blue
with aqueous solution of iodine. Sometimes a drop
of sulphuric acid is required to develop the reaction;
occasionally, however, the iodine gives a greenish or
brownish tint, owing to the admixture of nitrogenous
matters. These bodies undoubtedly belong to the
amylaceous, or starchy, group, and seem to resemble
cellulose in their general characters. They are not to
be confounded with lardacein, which was unfortunately
termed amyloid substance when first investigated,
from a misconception of its real nature.

(3) *Mucoid and colloid degeneration.*—These two
forms are often associated together, and at all times it
is difficult to draw a rigid distinction between them.
In mucoid degeneration the intercellular substance is
chiefly affected, in colloid degeneration the cells. In
the former there is an increase of mucin in the connective tissue elements, in the latter the gelatinous
constituents have apparently undergone changes, and
seem to be converted into collagen, and perhaps into
semi-glutin and hemi-collin. *Mucin* is found normally
in the adult in small amount in all connective tissues
proper; in the embryonic connective tissue it exists in
much larger quantities, the umbilical cord of fœtus
yielding it to a great extent. It is also derived from
the secretion exuded by epithelial cells (mucus), and
can be obtained in sufficient quantity for examination
from the saliva (§ 131, page 175) and bile (§ 139,
page 198). In normal urine it is present in small
quantities, which is much increased in catarrhal
conditions of the renal and urinary organs. Mucin is
soluble in alkaline solutions, from which it is precipitated by acetic acid. Mucin is not digested by
gastric juice, but is by the pancreatic ferment. It is
not precipitated by potassium ferrocyanide with

acetic acid, nor by tannic acid. Solutions of mucin with alkaline copper solutions prevents the precipitations of cupric hydrate, and no reduction takes place on boiling. *Collagen* is the name given to a substance obtained from the white fibres of connective tissue, and of which they are principally composed. It can be obtained tolerably pure by digesting finely sliced tendon in water for some days. Then filter off precipitate, and digest precipitate for some days in dilute baryta water; filter, and wash the insoluble residue thoroughly in water, then with dilute acetic acid, and again with water. Hofmeister has shown that when gelatin is heated for some time at 130° C. it loses water and becomes converted into a substance resembling gelatin; and also when collagen is boiled for more than twenty-four hours a further change occurs, the collagen taking up water and becoming converted into two peptone-like bodies, to which the names semi-glutin and hemi-glutin have been given. Hofmeister represents the changes by the following equations :

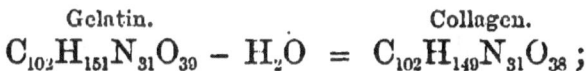

Gelatin.   Collagen.

$$C_{102}H_{151}N_{31}O_{39} - H_2O = C_{102}H_{149}N_{31}O_{38} ;$$

and

Collagen.   Semi-glutin.   Hemi-glutin.

$$C_{102}H_{149}N_{31}O_{39} + 3H_2O = C_{55}H_{85}N_{17}O_{22} + C_{47}H_{70}N_{14}O_{19}.$$

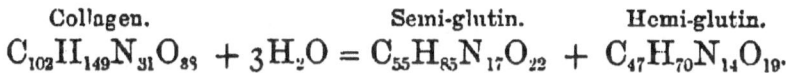

Professor Gamgee * states, that whilst mucin is unquestionably a product of the differentiation of the protoplasm of certain animal cells, and is obviously derived from the proteids, it is conceivable that it may result also from a decomposition in which collagen and mucin originate. What the nature of the decomposition may be is, however, quite unknown.

* "Physioloigcal Chemistry of the Animal Body," p. 259. Macmillan.

A further study of the subject is much to be desired,
as likely to throw considerable light on the pathology
of this degenerative process, especially with regard to
the chemistry of the contents of the ovarian cysts.
Professor Gamgee's account of these bodies is the best
summary extant of the work that has been done in this
direction. Still, there must be further investigation
in the physiological laboratory before the results can
be made available for clinical purposes. Anybody
taking up the subject from its present standpoint
would find a field for original research likely to
yield important results. At present, clinically, we
distinguish between mucoid and colloid degeneration
by the fact that in the former case the fluids are
precipitated by acetic acid ; in the latter, not. Mucoid
degeneration is by no means common, and seems to
imply a retrograde metamorphosis of the connective
tissue elements to their embryonic state. It occurs
occasionally in the intervertebral and costal cartilages
of old people ; it sometimes attacks the bones and
serous membranes. When localised, the softened
part surrounded by the firmer normal tissue gives
rise to a cyst-like formation. In an important com-
munication to the Royal Medical and Chirurgical
Society (*Trans.*, vol. lxi., p. 57) Dr. W. M. Ord,
describing the " cretinoid affection" occasionally
observed in middle-aged women, proposed that the
term *myxœdema* should be applied to express the
essential condition observable in that affection. That
condition relates to the jelly-like swelling of the
connective tissue, chiefly, if not entirely, consisting of
an overgrowth of the mucus-yielding cement by which
the fibrils of the white element are held together. In
one of Dr. Ord's cases the amount of mucin obtained
was more than a fiftieth of the quantity obtained from
the skin of non-œdematous bodies. The method
employed by Dr. Cranstoun Charles, who made the

chemical analysis for Dr. Ord, is given here, as it may be suitably employed for similar investigations. The skin of both feet was cut into pieces, and divided into three nearly equal portions, α, β, and γ. (α) was digested with water for several days; the filtrate from this was treated with an excess of acetic acid, let stand some hours, the precipitate separated on a filter and washed first with water acidified with acetic acid, and then with pure water. This washed precipitate was next left for twenty-four hours in lime-water, the solution filtered, and the precipitate again thrown down by excess of acetic acid. To purify the precipitate thus obtained it was washed successively with acidified water, pure water, alcohol, and ether, and then dried over a water-bath. The process is that employed by Eichwald for the separation of mucin from *Helix pomatia* and from tendons (*Ann. Chem. Pharm.*, Bd. 134, s. 177). (β) was left in methylated spirits for three days, in lime-water for two days, then filtered, and to the filtrate acetic acid added in excess. The precipitate was separated and purified as in α. (γ) was digested at once with dilute baryta water, the dissolved mucin precipitated by acetic acid, and purified as before. The body obtained by the above three methods was sensibly the same in each case in properties and in appearance, and nearly equal in amount; it corresponded in its reactions to the mucin of Scherer, Eichwald, and Staedeler. Mucoid degeneration also affects new formations; thus, enchondromatous, lipomatous, and sarcomatous tumours may undergo mucoid transformation, and may become wholly or partially converted into *myxomata*. In colloid degeneration the process commences apparently in cells, and not, as in mucoid degeneration, between the white fibres of the connective tissue element. The cells become filled with small masses of jelly-like material, which pushes aside the nucleus, and at length destroys the cell; after a time the intercellular substance atrophies and softens. In many cases a true mucoid degeneration of the intercellular substance seems to follow the colloid infiltration of the cells. This gives additional weight to Professor Gamgee's supposition that mucin and collagen originate from a decomposition common to both.

(4) *Calcareous degeneration*, or the infiltration of the tissues with calcareous particles. May be either general, as the result of an accumulation of calcareous salts in the blood—for instance, as in osteomalacia, where the earthy salts are removed from the bone, and are often in part deposited in other tissues, the lungs, stomach, or intestines. Or the infiltration may be local, due to changes in the tissue itself following some impairment of their nutrition; or as a consequence of a retardation or diminution of the amount of blood flowing through them, as I have endeavoured to show, may be one of the causes leading to calculous deposit in the tubules of the kidney (page 240). The cause of calcareous deposit in tissues has been explained to arise from the stagnation of the nutritive fluids, owing to which the free carbonic acid which holds the earthy salts in solution are precipitated, and, the circulation through the part being retarded, these insoluble elements are only partially removed, and so become deposited. The ossification of arteries proceeds from this cause. Calcareous deposits consist chiefly of calcium carbonate, calcium phosphate, magnesium phosphate, traces of soluble salts, and generally fatty matters and cholesterin. For their chemical examination, exhaust thoroughly with ether, to remove fatty matters, and proceed to estimate the earthy matters according to the directions given for the analysis of urinary calculi, Class II. (§ 152, page 248).

161. **Morbid exudations.**—(1) *Pus* is a pathological fluid, and consists essentially of a liquid
portion, "liquor puris," which is exuded liquor sanguinis, and white corpuscles, or leucocytes, which
cannot be distinguished from the white corpuscles of
the blood. The pus corpuscles can be separated from
the liquor puris by the addition of a 10 per cent.
solution of sodium chloride, and the precipitated
mass removed by filtration, and thoroughly washed
with the same solution till quite free from serum.
The pus corpuscles are spherical, irregular bodies,
about $\frac{1}{2500}$th to $\frac{1}{3500}$th of an inch, or 8μ to 10μ in
diameter, containing a number of granules and one
or more nuclei. When treated with dilute acetic acid,
they swell up, and become more transparent, and the
nuclei more distinct. Treated with ammonia or
potash solutions, pus becomes tenacious and jelly-like.
This character distinguishes it from mucus, which
becomes less tenacious and more fluid on the addition
of these solutions. According to Miescher, the nuclei
of the pus corpuscles contain a definite organic body,
containing phosphorus, which he calls nuclein. This
substance, he holds, can be obtained from all cells
where nuclei exist. It is, however, doubtful whether
it is a definite compound, since the results obtained by
analysis by different observers vary considerably. It
is more probable that it is an ordinary proteid substance, combined with lecithin or other phosphorised
bodies, in varying proportions. Eichwald has also found
peptone-like bodies in purulent fluids. The amount of
water and solids present in pus varies, of course, with
the nature of the pus. Thus, in ichorous, muco, or sero-
pus, the solids are diminished. The following is an
analysis of laudable pus, the result of acute inflammation : Water 87, Solids 13, in 100 ; proteids 8·5, fatty
matters 3·0, extractives 0·7, inorganic residue 0·8.
The proteids are sero-albumin and paraglobulin. The

s

fatty matters consist of neutral fats, cholesterin, and a body yielding glycerin phosphoric acid, probably lecithin, but some consider it to be protagon. The extractives contain traces of urea and glucose, sometimes leucin and tyrosin. A blue colour is often noticed on the dry bandages and linen which have been in contact with pus; this is due to *pyo-cyanin*. This substance can be obtained in a crystalline form by soaking the stained linen in water, containing a few drops of ammonia, for some hours, and then filtering off and evaporating the green liquid. The concentrated filtrate is then agitated with chloroform, and the chloroform solution removed and treated with very dilute sulphuric acid, and the mixture allowed to stand for some time. At length a red aqueous layer separates, which is removed and treated with caustic baryta till the solution becomes blue; filter, and agitate the filtrate with chloroform; remove the chloroform solution, and allow it to evaporate spontaneously, when pyo-cyanin will crystallise out in bluish-coloured needles or rectangular flakes. The crystals are soluble in water and chloroform, insoluble in ether. Acids turn their solutions red, but alkalies restore the blue colour; chlorine decolorises both solutions. The researches of Lücke and Fitz seem to prove that pyo-cyanin is generated by a bacillus, which, according to the latter observer, has the power of decomposing glycerin, in the presence of calcium carbonate, with the formation of carbonic acid, butyric acid, and hydrogen. It is probably the same bacillus that sometimes gives milk its blue colour.

From the chloroformic solution, after the removal of the pyo-cyanin, minute yellow crystals of pyo-xanthin can be obtained by evaporation. They are coloured red by acids, and violet by alkalies. Gelatin and chondrin are said to have been found in pus, but their existence as such is doubtful.

Perhaps the peptone-like bodies described by Eichwald may be related to semi-glutin and hemi-glutin (page 269). The inorganic residue consists chiefly of soluble salts of the alkaline chlorides, mostly as sodium chloride. The earthy salts in pus derived from the soft tissues are not abundant; but in pus derived from the neighbourhood of diseased bone, it may contain as much as 2·5 per cent. In making an analysis of pus, the paraglobulin is first precipitated by magnesium sulphate, and then the serum albumin coagulated by heat (§ 98, page 84). The liquid is then to be agitated with ether, to obtain the fatty matters (§ 99, page 85); and then the extractives are to be separated by the methods described for Urea (§ 100, page 86), Glucose (§ 100, page 88), Leucin (§ 169, page 126). A fresh portion of pus is then incinerated, and the bases and acids estimated according to directions at § 101, page 93. The specific gravity is taken with the specific gravity bottle (§ 92, page 66), and the reaction determined as directed § 93, page 67.

(2) *Dropsical fluids.*—Whenever the balance between the two processes, transfusion of the nutritive plasma from the blood-vessels and its re-absorption by the lymphatics, is disturbed, the quantity of parenchymatous fluid, which is always present in small quantity in the tissues, becomes increased. This disturbance of pressure of the fluid in the blood-vessels and that in the parenchyma may result from general or local conditions. In general dropsy, such as we find in certain diseases of the kidney, the condition depends on hydræmia, especially in those forms where the amount of water passing off through the kidneys is lessened. Thus, general dropsy is an important clinical symptom, distinguishing acute tubal nephritis, and the large white kidney from the granular and the albuminoid kidney, in both of which latter, large quantities of dilute urine are passed. In general, the

dropsy of kidney disease depends on the diminished
action of the kidneys, so that the water ingested
being the same, its exit is lessened; hence the
volume of blood being increased, the arterial pressure is augmented, and the serum passes into the
tissues in consequence of the increased pressure,
whilst its diluted condition perhaps renders it more
easy. In this way the whole of the tissues of the
body become more or less waterlogged. When the
dropsical effusion is localised, as in ascites, the exudation of serum depends on mechanical congestion due
to venous obstruction, and the increased pressure
takes place in the veins, and thus lessens absorption
by the lymphatics and veins. In the dropsy of heart
disease both forms are present. In diseases of the
mitral and tricuspid valves, which lead to dilatation of
the right side of the heart, obstruction of the venous
circulation occurs, and increased pressure in the veins,
which is at first most noticeable in those farthest
from the centre of circulation, and which from their
position (the lower extremities) allow the blood more
readily to stagnate in them. But diseases affecting
the right side of the heart also induce dropsy by
increasing the amount of water in the blood, since
their tendency is to diminish the flow of blood
through the kidney, and thus to diminish the
quantity of urine secreted. In disease of the aortic
valves, so long as compensatory hypertrophy is
maintained dropsy does not occur; but so soon as
the left ventricle fails, the rapidity of the circulation
through the kidneys falls, and the quantity of urine
excreted diminishes, so that the quantity of water
increases in the blood, and the condition of hydræmia
is induced. It will be thus seen that dropsical fluids
are blood serum more or less diluted with water.
This is an important point to remember; for if on
tapping a patient we find the amount of solids in the

fluids withdrawn in excess of those normally present, we may be sure that the fluid is not an ordinary dropsical effusion, and contains other matters besides blood serum. The specific gravity of a dropsical fluid ranges from 1·005 to 1·022, according to the amount of water present. The amount of proteids contained in it are variable, ranging from 0·4 to 6 per cent. They consist of ordinary sero-albumin, paraglobulin, and fibrinogen. Unless withdrawn from inflamed tissues, or blood is present, they rarely spontaneously coagulate, but do so on the addition of fibrin ferment, or are allowed to stand some time. They contain only a small proportion of fatty matters, about ·05 per cent., and in old-standing cases the ether extract yields cholesterin. The extractives are urea, a variable amount of glucose, and sometimes a little leucin. The spectrum of sero-lutein may be sometimes obtained by allowing the serum to stand and deposit, and then filtering it till quite clear. The clear fluid will give a band at F; and a very faint one between F and G (MacMunn). The salts are those of the blood. The exudation occurring in special serous cavities differs little from that poured into the subcutaneous areolar tissues. They are, perhaps, generally richer in proteid matters, and the fluid of the pericardium and hydrocele fluid yields a larger proportion of fibrinogen. The latter fluid also frequently contains succinic acid, and more cholesterin than is met with in other effusions. The cerebro-spinal fluid is accumulated in large quantities in cases of spina bifida, and chronic hydrocephalus. It is a clear fluid of low specific gravity, does not coagulate when heated, but becomes opalescent, and deposits a flocculent precipitate on the addition of acetic acid. It does not contain fibrinogen, so that it yields no coagulum when fibrin ferment is added to it. Sugar is said to be generally present, but some say that it is only to be found when there is irritation or inflammation of

the brain or spinal cord. It is to be regretted that
more complete analyses of this fluid have not been
made. It is a question that occasionally arises in
surgical practice (*Med. Clin. Trans.*, vol. ix., p. 247)
whether in punctured wounds of the back the aqueous
fluid that may be discharged from the wound comes
from the cerebro-spinal canal or from injury of the
kidney.*   In the case Mr. Holmes brought before the
Society it was doubtful whether the fluid came from a
wounded ureter or from the cerebro-spinal canal. An
analysis of the fluid which I made for Mr. Holmes
gave a very doubtful result. It was alkaline, and
contained 9·85 parts of water to ·15 of solids. The
solids consisted of an ordinary albumin coagulable by
heat, and an albumin not so coagulable, but precipitable by acetic acid ; there was a slight greasy residue
left from the evaporation of the etherial extract, but
no cholesterin. There were traces of urea, ·07 in 100
parts, but no uric acid or glucose. The salts consisted
chiefly of sodium chloride ·05 and phosphates ·03 in
100 parts. The amount of fluid that escaped from
the wound was considerable, three large draw-sheets
being soaked in the course of the day. The general
character of the fluid was not unlike that from a
spina bifida, except the coagulable albumin, which,
however, might have come from the wound. The
trace of urea was not more than one would expect
to meet with as an extractive in a fluid of this character, whilst it was greatly below the amount found
in even extremely dilute urine. On the other hand, it
is difficult to imagine such an enormous quantity could
have escaped from the cerebro-spinal canal for such a
continued period ; and again, it was noticed that when

---

* "Section of the renal nerves is followed by a most copious
secretion by what has been called hydruria and polyuria, the
urine at the same time frequently becoming albuminous."—
Prof. M. Foster, "Textbook of Physiology," p. 274.

the wound was partially closed the urine passed by the urethra increased in quantity.

The *liquor amnii* and *allantoic fluids* are usually clear and without colour. The proportion of water to the solids is very small. They contain traces of ordinary albumin, about 0·1 to 0·15 per cent., but contain no fibrino-plastic or fibrinogen. During the early period of gestation they contain a considerable amount of sugar, which, however, gradually disappears as pregnancy advances, till at the time of birth every trace has disappeared. (*See* page 206.) There is a variable quantity of urea always present. The allantoic fluid contains a characteristic ingredient *allantoin* (§ 67, page 50). The inorganic constituents of both fluids consist chiefly of sodium chloride and calcium and magnesium phosphate.

Synovia, or the secretion of the synovial membrane of joints, is denser than the fluid obtained from serous sacs, and more viscid from containing mucin. (For the systematic examination of dropsical fluids *see* Ovarian cysts, page 282.)

(3) *Contents of ovarian cysts.*—Considerable differences of opinion, especially as regards the chemical composition, exists as regards the nature of their contents. This is owing, no doubt, to the want of attention being paid to the nature of the degenerative processes that produce them. Thus, for instance, we have in some cases apparently only a simple fluid, poor in albumin, of low specific gravity, and which does not yield fibrin on the addition of fibrin ferment. In others the contents are viscid and jelly-like, rich in proteid elements, and which yield an abundant coagulum with fibrin ferment; in some of these, again, mucin will be found, in others it is absent, and we have instead a body resembling collagen or gelatin. It was formerly thought that paralbumin and metalbumin were characteristic

constituents of ovarian cysts, but they have been found in the contents of renal cysts. MacMunn believes that the spectroscope will enable us to distinguish the contents of parovarian and ovarian cysts from other fluids resembling them, and has described the spectrum in three cases of parovarian and one case of ovarian fluid that gave the same spectrum, that of acid hæmatin; viz., a band between c and d, nearer c, another between d and e; on adding ammonium sulphide to this fluid the bands of reduced hæmatin appeared at once. A fifth specimen, however, gave no spectrum whatever. At present, then, we have no one characteristic that can enable us chemically to distinguish between the fluid of an ovarian cyst and the contents of a cyst of another organ. If the solid constituents be above that of ordinary blood serum, we can say positively the fluid is not ascitic, but otherwise we have to depend on an examination of the whole constituents of the fluid before we can venture to form an opinion. The organic matters vary considerably from ·25 to 14·0 per cent.; the inorganic constituents, however, are tolerably constant, ranging from ·7 to ·9 per cent. The proteid substances consist of ordinary albumin, paraglobulin, sometimes of fibrinogen. Two proteid bodies paralbumin and metalbumin, are so generally present they are supposed to be characteristic constituents. Their composition seems to be doubtful. Paralbumin was first obtained by Scherer; its alkaline solutions are ropy and viscid; it is precipitated from its warm solutions by carbonic acid gas, but not by magnesium sulphate; it is coagulated by nitric acid, but the precipitate re-dissolves in strong acetic acid. It seems to be associated with a body resembling glycogen, and capable of being converted into a substance giving the reactions of dextrose with copper. Paralbumin may be a transitional form of mucin, since Eichwald has found that when heated with dilute mineral acids, mucin yields acid albumin, and another body which closely resembles dextrose by reducing solutions of cupric sulphate. Metalbumin closely resembles paralbumin, and perhaps is more closely related in its reactions to mucin than that body. Both bodies may therefore be intermediate products of the transformation of proteid substances into mucoid or colloid matter. Mucin is found in some cysts, but not in all; it is the most variable of all the organic constituents. Ovarian fluids yield about 0·6 per cent. of fatty matter, and sometimes crystals of cholesterin separate out and float on the surface. The salts do not differ materially from those of blood-ash. The fluid from *hydatid cysts*, if the sac be not inflamed, is limpid when running from the cunnula, but becomes opalescent on standing. The reaction is alkaline, and the specific gravity ranges from 1·009 to 1·013. According to Murchison, the fluid of the hydatid cysts contains no albumin, but Naunyn has found traces. If to the contents of an hydatid cyst magnesium sulphate be added, and then a stream of CO_2 be passed through, and the fluid heated to 75° C., a fine precipitate of paraglobulin and seroalbumin will generally be obtained if there has been inflammation of wall of the cyst. The fluid, however, contains no urea. The great characteristic, however, of the fluid of an hydatid cyst is the considerable amount of sodium chloride it contains, a quantity not found in any other fluid in the body whether healthy or morbid. These characters, even if the "hooklets" are absent, are generally sufficient to determine the nature of the fluid. It may, however, be mistaken for the fluid from *hydro-nephrotic cysts*. This is of low specific gravity, 1·004, contains no albumin (paralbumin and cholesterin have been found by Dr. Schetelig, of Hamburg, in a renal cyst).

Urea and uric acid are often absent; the reaction is generally faintly acid, more frequently neutral, and although containing an abundance of sodium chloride, it is not so rich in this salt as the fluid of an hydatid cyst. The method of procedure for an analysis of dropsical or ovarian fluids is as follows : Evaporate a weighed portion of the fluid (§ 92, page 66) to ascertain proportion of water and solids, and the residue incinerated to determine the saline constituents (§ 101, page 93); the reaction is ascertained (§ 93, page 67); the proteids are determined by precipitating from a weighed quantity of fluid the paraglobulin, and fibrinogen if present, by precipitation with magnesium sulphate (page 34); remove precipitate, and acidulate filtrate with a few drops of dilute acetic acid, and coagulate the ordinary albumin by heat (§ 98, page 84). After the removal of the proteids, evaporate the filtrate to dryness, and extract with ether to remove fatty matter, which can be estimated according to directions given, § 99, page 85. The residue, after the removal of the ether, is then treated successively with boiling water and boiling alcohol, and the aqueous and alcoholic extracts mixed together and evaporated to dryness, and then dissolved in hot water, and the amounts of urea and glucose determined by processes given at § 100, pages 87 and 88. In a fresh sample of the fluid examine for special products, as paralbumin, metalbumin, succinic acid ; take the specific gravity according to directions § 92, page 65 ; examine fluid with spectroscope for bands sero-lutein (page 277), or hæmatin (page 77).

162. **Chemical changes in bone** chiefly relate to the variations occurring between the insoluble salts in relation to the organic basis. Thus, for instance, in the atrophy of bones met with in cases of long-standing anchylosis, dislocations, etc., when the compact and cancellous tissue gradually becomes

absorbed, although the whole bone is smaller and lighter, yet the proportion of inorganic constituents to the organic is relatively more than in healthy bone. The same occurs as the result of chronic inflammation of bone, where new formation of osseous tissue takes place in the enlarged Haversian central cancellous spaces, and the whole bone is converted into a dense, heavy mass. In rickets and osteo-malacia, the opposite condition maintains, and the bones affected have their calcareous elements diminished. The following tables, giving analyses of the tibia, will best show the composition of bone under different conditions :—

ANALYSES OF NORMAL BONE (TIBIA).

	Fœtus. 7 months.	Child. 1 year.	Child. 5 years.	Man. 25 years.	Cortical portion.	Medullary portion.	Head of the bone.
Organic matters	40·37	43·42	32·29	31·97	38·02	41·16	48·56
Calcium phosphate	53·46	48·55	59·74	58·95	52·93	49·01	41·77
Magnesium phosphate	2·00	1·00	1·34	1·30	0·25	1·54	0·87
Calcium carbonate	3·10	5·79	6·00	7·08	7·66	7·76	7·10
Soluble salts	1·07	1·24	0·63	1·55	1·11	0·52	1·67

ANALYSES OF MORBID BONE.*

	Rickets. Lehmann.	Osteomalacia. Lehmann.	Caries. Vallentin.	Osteo-sclerosis. Ragsky.	Exostosis. Von Bibra.
Organic matter	66·36	75·69	55·88	44·10	48·55
Calcium phosphate	26·94	18·83	34·38 ⎫	48·20	⎧ 47·99
Magnesium phosphate	0·81	0·54	1·18 ⎭		⎩ 1·55
Calcium carbonate	4·88	3·83	6·63	7·45	1·00
Soluble salts	1·08	0·43	1·91	0·55	0·91

* These analyses were made from typical, but not extreme, instances of the disease.

Bone contains less water in proportion to its solids than any other tissue, except enamel and dentine, pulverised bone evaporated to dryness losing only 10 per cent. of its weight. From the fact that when perfectly dried bone is placed in water there is a slight elevation of temperature, it is supposed that the water exists in bone partly in a state of chemical combination, like the water of crystallisation. The organic matter may be obtained by digesting a fragment of bone for some time in dilute hydrochloric acid (1—5), when the inorganic matters will be dissolved, leaving the gelatinous matrix intact. This consists of a body identical with collagen (which, by long boiling, yields gelatin, and gelatin again, by being heated above boiling-point (130° C.), is reconverted into collagen) mixed with a small quantity of elasticin, and traces of proteid, and some fat. Most of the fat of bones, however, is found in a free state in the yellow marrow of the medullary cavity of the long bones, and, to a lesser extent, in the red marrow of the cancellated tissue. The yellow marrow differs little from ordinary fatty matter, and consists chiefly of an admixture of neutral fats. The red marrow yields much less fatty matter than the yellow; it contains a proteid body; a free acid (lactic acid?), a product of decomposition; hypoxanthin; and certain granules which stain a deep blue with potassium ferrocyanide, and therefore probably contain iron. In addition, there are large multi-nucleated cells (*Myeloplaxes of Robin*), which play an important part in the absorption and formation of bone. These and the granular bodies, according to Heitzmann, Malassez, and others, have also to do with the formation of the red blood-corpuscles. In pernicious anæmia and leucæmia, where nucleated coloured corpuscles similar to those found in the blood of the human embryo exist, the marrow of the bones is also generally affected ; hence the term "myologenic leucæmia." The inorganic constituents can be obtained by incinerating some of the crushed bone in a muffle furnace, which burns off the organic matter. In addition to phosphoric acid, carbonic acid, and chlorine in combination with lime, there exists a minute quantity of fluorine, about 0·22 parts in 100 of bone-ash.   Some of the other metals are occasionally met with, as strontium, aluminum, silicon, copper, and arsenic.   By long keeping, bones yield relatively less organic matter than fresh, but their inorganic constituents are little altered.   Dr. Norman Moore has (*Path. Soc. Trans.*, 1883) shown some interesting specimens of fossil bone, which were evidently from a subject that had suffered from chronic rheumatic arthritis.

To make an analysis of diseased bone, it is advisable to take a section representing both the compact and cancellated part, and a similar section in which the compact part is carefully divided from the cancellous, and submit them all to a separate analysis. In this we learn the chemical changes that have taken place in the bone as a whole, and the changes that have affected the compact and cancellous portions respectively.   If possible, it is as well, from the same subject, to examine a portion of a bone as nearly resembling in form and structure the one diseased, but which is apparently free from the disease.   The fragments of bone are to be reduced to a fine powder, and the weight ascertained.   (Ten grammes is a convenient quantity to take.)   The powder is then placed in a weighed platinum crucible, and dried on a calcium chloride bath at a temperature of 130° C. till it ceases to lose weight.   The loss of weight represents the amount of water.   The finely-dried powder is extracted with ether, by means of the apparatus described page 263 ; this gives the amount of fatty matter; and the etherial residue is examined, as directed § 99, page 85, to determine the nature of the fats. The powder, after exhaustion by ether, is again weighed in the weighed platinum crucible, then incinerated in a muffle furnace till the ash is quite white, allowed to cool over sulphuric acid, and, when cold, weighed. The loss of weight represents the organic matter; the actual weight, the inorganic residue. The ash is then boiled at 100° C. in distilled water for some time, by which means the soluble salts are dissolved. Filter, evaporate filtrate, and weigh; the weight of residue represents weight of the soluble salts. If required, estimate the sodium by process described § 88, page 61, and the hydrochloric acid, § 114, page 136, and the carbonic acid, page 287. The insoluble residue not dissolved by boiling water is then dried and weighed, the weight representing the insoluble salts. Dissolve these in acetic acid, and divide the solution into two equal parts. One portion of the solution is then examined according to the directions given §§ 84, 85, for quantitative determination of lime and magnesia. The other portion is heated according to the directions given for the estimation of phosphoric acid § 113, and for hydrochloric acid, § 114. Having obtained the total amounts of sodium, calcium, and magnesium, and the amounts of phosphoric acid and hydrochloric acid, the result of the analysis is calculated as follows: The amount of hydrochloric acid obtained by the aqueous solution is combined with the sodium, and calculated as sodium chloride, and any overplus of soda is combined with the carbonic acid and reckoned as sodium carbonate. The whole of the magnesium is combined with phosphoric acid as magnesium phosphate, and the excess of phosphoric acid is deducted from the original amount of phosphoric acid obtained by volumetric analysis. This is combined with the lime, and calculated as

calcium phosphate. But as it is insufficient to combine with the whole of the lime, the amount of hydrochloric acid obtained from the acid solution (not from
the aqueous solution, which has been combined with
sodium to form sodium chloride) is calculated as
calcium chloride, and the portion of the lime that
still remains uncombined is then taken as calcium
carbonate. The fluorine may be determined either
by making an exact determination of the carbonic
acid existing in lime, instead of assuming that the
whole of lime that is not combined with phosphoric
acid and hydrochloric acid as given above. It is then
found that a minute quantity of lime is in excess of
what is required to form calcium carbonate, and this
quantity represents the amount that combines to form
calcium fluoride. To estimate the quantity of carbonate lime, a weighed quantity of unburnt bone,
finely pulverised, is to be thoroughly extracted with
boiling water to remove soluble chlorides. The
residue dried, and placed in a glass flask fitted with a
desiccating tube ; a test-tube containing hydrochloric
acid is then placed in the flask, so arranged that by
means of a fine wire passed through the cork of the
flask it may be allowed to flow gradually over the
dried portion of bone at the bottom of the flask. The
apparatus and its contents are then weighed ; the
hydrochloric acid is poured gently over the powdered
bone, till it ceases to effervesce on addition of acid.
The flask is then shaken from time to time to dis-
engage any gas that may be remaining, and afterwards
weighed, the loss of weight representing the amount
of carbonic acid in combination with the lime. Now,
by the previous process we have determined the total
amount of lime in the ash of bone, and have found
how much combines with the phosphoric acid, how
much with hydrochloric acid, and now we have found
the quantity that goes to form calcium carbonate ; the

overplus of lime, therefore, that remains after these calculations may be reckoned as calcium fluoride. The determination, however, of calcium fluoride is of little practical value in clinical or pathological investigations, so that the direct determination of the calcium carbonate need not necessarily be resorted to ; and the analysis of bone may be concluded by the indirect determination of the carbonate, that is, by the process of calculating out first the amount of lime combined with phosphoric acid, and the amount combined with hydrochloric acid, leaving the overplus of lime to be calculated as carbonate. The presence of fluorine can be determined by treating large quantities of the ash of bones with strong sulphuric acid, and gently heating the mixture in a glass vessel, when the glass will become corroded. If the glass vessel has been previously weighed, the amount of fluorine can be determined by the loss of weight the glass sustains.

With regard to the chemical changes taking place in bone in rickets, much difference of opinion has been expressed. It has been held that the calcareous salts are not retained, owing to the inability of the organic matrix to fix them ; or that they are dissolved out of the matrix, owing to the excessive formation of lactic acid in the system. These views have been based upon the supposition that the urine contains an excess of lime salts. This is based, however, on very insufficient evidence, and is not confirmed by the more recent analyses of urine from rachitic patients, which points very conclusively to the fact that the lime salts are not excessively excreted by the kidneys in this disease. Dr. Seeman (*Zur Pathogenese und Etiologie der Rachitis; Virchow's Archiv.*, lxxvii., 1879), who has analysed the urine of sixteen rachitic children, has actually found a diminution, which was most marked when the disease was at its height. It is probable that the bone changes in rickets are due to a lime starvation rather than to a lime withdrawal; the salts of that base not being introduced in sufficient quantity into the system, the catarrhal condition of the mucous membrane of the intestines, which invariably accompanies this disease, hindering their absorption. A condition termed hæmorrhagic rickets has lately been described; it is doubtful whether this is related in any way to scurvy, or, to speak more accurately, the result of a scorbutic condition, or is merely an exaggeration of the hyperæmic condition of the periosteum, which is more or less always noticeable in the bones of rickety children. This is a question for histologists to decide, but it is an interesting fact that these cases improve under the administration of lime-juice. In osteomalacia the poverty of bone in calcareous constituents evidently depends on the reabsorption of the lime salts, since not only does the urine demonstrably show an excess of earthy phosphates, but these salts are also deposited in the tissues. (*See* Calcareous degeneration, page 272.) Some have supposed that the withdrawal is due either to excessive formation of lactic acid in the system, or to its local formation in the bone itself. The fact that lactic acid administered in large quantities, as in the lactic acid treatment of diabetes, or experimentally in the case of animals, seems to show that excessive formation of lactic acid is not the cause. It is probable in this case that the process is truly a degenerative one, like myxœdema, and in which the organic matter of the bone reverts to its embryonic condition, which Morochowitz believes to be a mixture of collagen and mucin. It is interesting to observe, in connection with its probable relation to myxœdema, that it most frequently attacks women, at about the same age when myxœdema makes its appearance.

T

Scurvy, gout, and rheumatism. — These diseases may be considered as typical examples of disorders arising, in the first instance, from chemical alterations in the quality of the blood. In their early stages, or ill-defined forms, they often present many features common to each other. Indeed, the close resemblance between gout, rheumatism, and scurvy in their early stages repeatedly attracted the notice of the older writers on the subject. Thus, Sydenham* says, "where matter suited to produce the gout is newly generated there appear various symptoms which occasion us to suspect the scurvy, till the formation and actual appearance of the gout remove all doubt concerning the disorder." The symptoms common to these disorders in their early stages may be thus briefly enumerated. Fugitive and erratic pains in the limbs; tenderness of the joints; attacks of dyspnœa, more or less paroxysmal in character; severe attacks of pain over the region of the heart; weak and intermittent pulse; irregular discharge of urine, sometimes profuse and of low specific gravity, at other times scanty and concentrated; and a tendency to subcutaneous hæmorrhages; whilst one point in common to these symptoms, worthy of especial consideration, is their extreme motility, the paroxysmal nature of their onset, the suddenness with which they disappear or transfer themselves from one region or organ to another. These sudden changes afford additional support to the view that these derangements are caused by chemical alterations in the quality of the blood. What the precise nature of these chemical changes may be is not yet definitely established, but evidence is gradually increasing which points more and more to the conclusion that they arise from a diminution of the normal alkaline reaction of the blood.

* "De Rheumatismo," sect. 6, cap. v.

With regard to scurvy, the most important observations have been made by Mr. Busk and Dr.
Garrod. The former, in a series of analyses of blood,
published in Dr. G. Budd's article on Scurvy, in the
"Library of Medicine," showed that in this disease there
was a considerable diminution of the corpuscles and
increase in the fibrin, and an augmentation of the inorganic residue. Unfortunately, Mr. Busk did not
complete his observations by making a separate estimation of each of the inorganic constituents. In 1848,
however, Dr. Garrod found that in the urine of scurvy
patients the potash salts were considerably diminished.
This observation of Dr. Garrod I was able to confirm
in 1877.* Dr. Garrod thought that this diminution
of potash was due to a deficiency of this base in the
food of those affected, but it has been subsequently
shown that this cannot be the case, since peas, which
form an important item of the sailor's diet, are rich in
this substance ; and, moreover, if we give patients
suffering from the disease unlimited quantities of
Liebig's extract of meat, a food extremely rich in
potash salts, no amelioration occurs till lemon juice,
potatoes, or other fresh vegetable substance is administered as well. This suggests that scurvy does
not depend on the withdrawal of potash from the
system, but on some alteration subsisting between the
organic acids and the base. With regard to this point,
I was able to establish the important fact that the
urine passed by scorbutic patients was deficient not
only in potash but also in the alkaline phosphates.
Thus, after the withdrawal of all vegetable food for
eighteen days, I found the alkaline phosphates had
sunk from 2·1 grms. to 1·5 grms., whilst there was a
slight increase of the earthy phosphates. Again, in
four cases of scurvy, two of which are related in my

* "An Enquiry into the General Pathology of Scurvy." Lewis,
1877.

pamphlet, and two I have subsequently observed, I found in the first case that the alkaline phosphates on admission were as low as 0·76 grm., but that on the administration of lemon juice, the diet in other respects being the same, they rose to 1·6 grms. In the second case, the amount on admission was 0·57 grm., which rose after administration of lemon juice to 1·6 grms. In the third case the rise was not so marked, being from 1·25 grms. to 1·77 grms.; and in the fourth case the lemon juice increased the amount from 0·87 grm. to 1·29 grms. In all these cases the diet was the same throughout, the only difference being the administration of a small quantity of lemon juice (2 oz. daily), which could not possibly account for the decided increase of the alkaline phosphates. It therefore seems to me that these salts are retained in the system in scurvy to supply the deficiency of the other alkaline salts, the alkaline carbonates and bicarbonates, which are withdrawn when fresh vegetables are withheld. This view is in accord with the experiments of Hoffmann and Loscar, to which attention has been already called (§ 93, page 68). There can be no doubt that the alkalescence of the blood is mainly dependent on the formation of the alkaline carbonates derived from the reduction of the organic salts, the lactates, malates, citrates, etc., introduced with the food, chiefly by vegetables, but also, as the experience of Dr. Neal in the Eira Expedition tends to show (*Med. and Chir. Soc. Trans.*, 1883), with fresh meat as well. When this supply is cut off, as is the case in long voyages or in sieges, etc., the blood is deprived of one of its sources for maintaining its alkalescence at its normal point, and consequently it falls back on the other alkaline salts, viz., the alkaline phosphates, to supply the deficiency. In this way it may maintain the proper degree of alkalinity for a time, but after a while a minimum point is reached which is incompatible with healthy nutrition, and textural changes are the result. These changes in scurvy resemble closely those induced in animals, in whom, by feeding with food rich in acid salts, or by the direct administration of acids, the alkalinity of the blood has been slowly diminished, viz., dissolution of the blood globules, ecchymoses in the heart, blood-stains in the mediastinum, gums, and mucous surfaces; whilst the muscular structure of the heart and muscles generally, as well as the secreting cells of the liver and kidney, become granular and even distinctly fatty (§ 93, page 69). With regard to the influence of the organic salts in the prevention of scurvy, it is interesting to remark that their range is apparently limited. Experience has shown us that fresh vegetables alone are efficacious as anti-scorbutics, and the same may be said of meat. In Arctic regions, where the meat is frozen almost as soon as killed, and in hot countries, where it is eaten before the changes induced by rigor-mortis set in, meat is considered as highly anti-scorbutic. In Europe, where the meat is generally kept some time before it is used as food, it is not so reputed. The explanation of this lies in the fact that fresh vege-tables and meat undergo changes by keeping with regard to the relation of the organic acids to the bases. Thus, the juices of fresh vegetables contain a certain quantity of organic acid (malic, citric, tartaric, etc.) in combination with a definite quantity of a base. On keeping, fermentative changes take place in the saccharine elements (as we see in fodder preserved by ensilage), and an additional quantity of free organic acid is formed (this in the case of vegetables is acetic acid), whilst the quantity of the base in the vegetable remains constant. Now this additional acid takes a portion of the base from the existing salts, reducing them from alkaline or neutral salts, with three or two equivalents of base, to acid salts, with only one equiva-lent. These acid salts, the malates, citrates, acetates, etc., in turn are reduced in the blood to acid carbonates (bicarbonates), salts which, though they may have an alkaline 'reaction, play the part of an acid in the system, and effect decompositions with neutral salts in all respects like acids (see formulæ, pages 24, 70, 183). The same may be said with regard to meat, since after rigor mortis* has set in there is an increase of lactic acid in its juices, so that acid instead of alkaline lactates are formed. From the foregoing it will be gathered that the development of scurvy depends in the main on a change in the conditions of the blood, in the direction of a diminution of its alkalescence, and that this diminution may be caused either by the direct cutting off of organic salts which yield by oxydation alkaline carbonates in the blood ; or indirectly, though not nearly so immediately, by the continued use of food which contains acid instead of alkaline salts. And further, that this change in the composition of the blood is more real than apparent, since the acid salt is a bicarbonate of potash or soda, a salt which has an alkaline reaction, and so apparently continues to give blood its alkaline reaction, whilst its chemical constitution is that of an acid salt, and acts as such in the chemical decompositions it occasions in the body. It is only by this means, as we have seen when discussing the nature of the reaction of the urine and the gastric juice (§§ 107, 179), that we are able to account for the seeming paradox of the separation of these acid secretions from the alkaline blood.

* It has been pointed out as an objection to the view that kept meat has no anti-scorbutic virtues, that the South American hunters subsist almost entirely on dried hung meat. So that if scurvy depended on the undue development of acid in the juices of the meat after having been killed some time, these men ought to suffer from scurvy, whereas they are remarkably free from the disease. It is probable, however, that the rapid drying of the freshly killed meat under a tropical sun puts a stop to the fermentative changes that produce excess of lactic acid, so that the "jerked" beef of tropical countries resembles the frozen meat of Arctic regions.

In gout a diminution of the alkaline reaction of
the blood has also been observed. Dr. Garrod has
pointed out, that with the exception of collapsed
cholera, and perhaps certain cases of albuminuria, the
reaction of the blood is to be found nearer the neutral
point in severe forms of chronic gout than in any other
disease. This diminished alkalinity in the blood of
gout has a relation to that which happens in scurvy,
for though it does not depend on the actual withdrawal
of alkaline salts supplied by fresh fruits and vegetables,
yet the diminution is caused by the addition of acids
or acid salts taken in excess with the food, or retained
in the system, the result of imperfect elimination, etc.
It is generally held that the excessive formation of uric
acid is the reason of the phenomena observed in this
disease, but the idea is gaining ground that the presence of this substance in the blood, and its deposit in
the tissues, is not caused by excessive formation, but to
the fact of its retention in the system and to its insolubility. When speaking of uric acid (§ 111) it was
stated that this substance is never found in normal
blood, or perhaps even in the blood of any disease
except gout, and it is considered probable that in
man it is not formed in the body to the extent that
had been supposed. Indeed, the evidence seems to
point to that fact, that the quantity formed in health
is extremely small, and that it is destroyed almost as
soon as formed, and so never enters the general current
of the circulation. Whilst the small quantity (0·5 grm.)
that escapes daily by the kidney does not probably represent the amount formed in the body, but is simply the
uric acid formed by the kidney itself, and which passes
directly out of the body instead of being destroyed. In
gout, however, we have decided evidence that uric acid
is present in the blood, that it is deposited in certain
tissues; and also, although the evidence is not quite
satisfactory to me that even the small quantity of uric

acid which appears in normal urine is reduced. Dr.
Garrod considers gout to depend upon a failure of the
renal function to excrete uric acid. In this way he
accounts for the diminution of uric acid from the urine
and its accumulation in the blood ; whilst its deposition
in the tissues depends on this accumulation, and on the
fact that uric acid is present in the blood in the form
of insoluble urate of soda. If it could be satisfactorily
proved that the minute quantity of uric acid, 0·5 grm.,
constantly present in healthy urine, really came from
the blood, or that it was proved beyond doubt that this
extremely minute quantity was still further invariably
reduced in gouty patients, Dr. Garrod's view might be
accepted without challenge. But there are difficulties
in the way of accepting the belief that the uric acid of
the urine comes from the system generally, whilst its
diminution in gout is by no means invariable ; for
where this diminution is chiefly observed is in chronic
cases, where kidney changes have been established.
In fact, I venture to think the accumulation of uric
acid in the blood, and its deposit in certain tissues,
depend on other conditions than failure of the renal
function, and that the first step in the process lies in
the failure of the tissues to reduce the uric acid formed
in them, as is the case in health. In the large glands,
or where the current of the circulation is free, the uric
acid is carried into the blood and gradually reduced to
urea ; in tissues outside the current of the circulation
the insoluble uric acid is not so readily carried off,
and so on the slightest disturbance is deposited, as is
the case in cartilages of the joints, the ear, etc. The
conditions which prevent the normal destruction of
uric acid in the tissues, and which permit it to pass
into the circulation, depend probably on disturbance of
innervation. These conditions I have endeavoured to
formulate in the chapter on the derangements associated with deposits of uric acid, in my work " On

Morbid Conditions of the Urine" (p. 65). As they are too lengthy to dwell upon here, it will be sufficient briefly to state the conclusions arrived at, which are that the first step in the pathology of gout is a textural degeneration, either hereditary or acquired, by which the tissues and blood become loaded with effete products; that such *predisposing* conditions lead at last to a disturbance of some special trophic nerve centre, caused either by a degenerative change in its structure, or derangement of its functions by the circulation through it of impure blood. This disturbance may be considered the *determining* cause of the gouty attack. The *result* is the accumulation of uric acid in the blood, and the deposition of urate of soda in the tissues.

With regard to the diminished alkalinity of the blood noticed in chronic gout, although no doubt in some measure due to the presence of uric acid and acid urates, yet excess of other organic acids undoubtedly has to be taken into consideration. For if the acids concerned in the production of gout were derived solely from the nitrogenous elements of the food and tissues, then by a rigid limitation of animal food within physiological limits we might hope to check or control the progress of the disease. But other articles of diet besides the nitrogenous, or those which, like alcohol, disturb the functions of the liver, and are thus supposed to lead to increased formation of uric acid, give rise to gouty symptoms; and it is a common experience with the gouty that there is as much arthritic trouble in a plateful of apple tart as in a mutton chop, and in a few strawberries as in a glass of port wine.

In rheumatism, no observations, as far as I am aware, have been made to determine whether there is a diminution of the alkaline reaction of the blood ; but no one can have observed the enormous quantity of acid poured out from the body by the skin in this disease, or have noted the high degree of the acidity of the urine during the progress of the attack, without coming to the conclusion that an excessive formation of acid is going on in the system. What the nature and character of the acid is, or how formed, we know absolutely nothing, though some have supposed it to be lactic acid. Whatever the acid may be, its development seems to be local rather than general, and is apparently excited by catarrhal influences rather than by previous accumulation of acid in the tissues and fluids. The acute manifestation of the disease occurring for the most part among young adults, or during the earlier period of middle age, there is not the same impairment of tissue as is the case with gouty patients, which may account for the non-deposition of urate of soda; or else the specific character of the inflammation being excited by a different cause, and not to a prolonged saturation of the tissues with acid products, does not lead to the accumulation and deposit of uric acid in the tissues.

INDEX.

————◆c◆————

U

CASSELL AND COMPANY, LIMITED, BELLE SAUVAGE WORKS, LONDON, E.C.

www.ingramcontent.com/pod-product-compliance
Lightning Source LLC
Chambersburg PA
CBHW021503210326

41599CB00012B/1118